FROM EVE TO EVOLUTION

# FROM EVE TO EVOLUTION

Darwin, Science, and Women's Rights
in Gilded Age America

KIMBERLY A. HAMLIN

THE UNIVERSITY OF CHICAGO PRESS

CHICAGO AND LONDON

KIMBERLY A. HAMLIN is associate professor of American studies and history at Miami University in Oxford, Ohio.

The University of Chicago Press, Chicago 60637
The University of Chicago Press, Ltd., London
© 2014 by The University of Chicago
All rights reserved. Published 2014.
Printed in the United States of America

23 22 21 20 19 18 17 16 15 14    1 2 3 4 5

ISBN-13: 978-0-226-13461-1 (cloth)
ISBN-13: 978-0-226-13475-8 (e-book)
DOI: 10.7208/chicago/ 9780226134758.001.0001

Library of Congress Cataloging-in-Publication Data
Hamlin, Kimberly A., author.
    From Eve to evolution : Darwin, science, and women's rights in Gilded Age America / Kimberly A. Hamlin.
        pages ; cm
    Includes bibliographical references and index.
    ISBN 978-0-226-13461-1 (cloth : alkaline paper) — ISBN 978-0-226-13475-8 (e-book)    1. Feminism and science—United States—History—19th century. 2. Evolution (Biology) and the social sciences—History—19th century. 3. Women's rights—United States—History—19th century.    I. Title.
    HQ111.U6H36 2014
    305.420973090'34—dc23

                                                        2013037257

♾ This paper meets the requirements of ANSI/NISO Z39.48-1992 (Permanence of Paper).

TO MY HUSBAND, MICHAEL,

AND OUR DAUGHTER, RUBY

# CONTENTS

# Evolution and the Natural Order

In 1913, the bohemian journalist Floyd Dell declared "the woman's movement is a product of the evolutionary science of the nineteenth century. Women's rebellions there have been before. . . . But it is modern science which, by giving us a new view of the body, its functions, its needs, its claims upon the world, has laid the basis for a successful feminist movement."[1] To modern readers, this may seem a curious statement. Prominent women's suffragists did not invoke evolutionary theory very often, and Charles Darwin's ideas about women, like those of most of his fellow evolutionists, were largely shaped by the ideology of "separate spheres" for men and women that dominated the Victorian era in which he lived. Visionary scientist, yes; feminist, no. Furthermore, since the 1970s, feminist historians have frequently argued that Darwinian evolutionary science, at least as it was articulated in the nineteenth century, should be considered, in the words of one scholar, "intrinsically anti-feminist."[2] What then could Floyd Dell have meant? This book suggests that Charles Darwin (1809–1882), the most influential evolutionist of the nineteenth (or any) century, did not intentionally upend traditional ideas about gender and sex, but that is precisely what his writings helped to do, as many American women's rights activists immediately recognized.

For generations, all one needed to know about the proper place of women in American and European society could be gleaned from reading Genesis, which explained that woman was created from man's rib to be his helpmeet, only to introduce sin into the world and cause the fall of mankind. When women imagined themselves in the world they thought in terms of Eve, and when men had to pinpoint why women could not attend university, minister in church, or participate in public life, they, too, drew on the story of Eve. Adam and Eve provided the script, the images, and

the template for Western ideas about gender until Darwinian evolutionary theory challenged their very existence and made it possible for women
and men to imagine alternative origins and a whole new range of gendered possibilities. Furthermore, at exactly the same time that Americans
were grappling with evolutionary theory, the burgeoning women's rights
movement brought questions of sex difference to the forefront of public
debate, making feminism and evolutionary theory concurrent intellectual
developments in the transition from the Victorian to the modern era. This
historical confluence of events also meant that women and men alike often sought answers to "the woman question" in evolutionary theory.

From Eve to Evolution analyzes American responses to evolutionary
theory through the lens of gender, and it provides the first book-length
study focusing on nineteenth-century women's responses to evolutionary theory. The following chapters pay particular attention to the women,
and a few men, who sought to combine their enthusiasm for evolutionary
theory with their commitment to women's rights, individuals who might
best be grouped under historian Beryl Satter's concept of "reform Darwinists." Reform Darwinists defined themselves as progressive evolutionists,
in favor of things such as worker's and women's rights and in opposition
to social Darwinists, who tended to support Gilded Age industrial inequi-
ties and the status quo.[3] These women, including Antoinette Brown Blackwell, Helen Hamilton Gardener, Eliza Burt Gamble, and Charlotte Perkins
Gilman, forged an evolutionary feminism that grappled with questions of
biological sex difference, the extent to which maternity did (and should)
define women's lives, the equitable division of household labor, and female
reproductive autonomy. The practical applications of this evolutionary
feminism came to fruition in the early thinking and writing of the American birth control pioneer Margaret Sanger. Much has been written about
what Darwin and other male evolutionists had to say regarding women;
little has been written about what women had to say regarding evolution.[4]
This project is one attempt to add women's voices and a focus on gender to
the vast literature on Darwin in America.

Darwin's own views on gender, at least as expressed in his published
writings, often rearticulated the dominant, patriarchal views of his era.[5]
In the nineteenth century, prescriptive literature and social customs dictated that men inhabit the worlds of commerce, labor, and politics, while
women controlled the home, the family's spiritual life, and the children.
Such a gendered division of labor was considered natural, civilized, and in
accordance with God's will, and, at first glance, Darwin's writings about
evolution did little to challenge these long-standing beliefs. Darwin's ideas

regarding sex differences will be explored more fully in subsequent chapters, but, in brief, he explained that, throughout the animal kingdom, the male "has been the more modified" due to the males' having "stronger passion than the females," which tend to retain "a closer resemblance" to the young.[6] Among humans, Darwin believed that "owing to her maternal instincts" woman differs from man chiefly in her "greater tenderness and less selfishness" and lack of intellectual attainments.[7] Overall, Darwin believed that female intellectual inferiority was natural and, most likely, immutable; he imposed Victorian gender roles and mating behavior on animals—combative male insects, strutting peacocks, and coy peahens; and he espoused patriarchal marriage as the epitome of civilization. Such descriptions inspired at least one generation of naturalists to conclude that women's inferiority was a permanent and necessary part of the evolutionary process and a later generation of feminists to reason that evolutionary science was inherently misogynistic.

Yet, Darwin's writings, especially *The Descent of Man, and Selection in Relation to Sex* (1871) and its cornerstone theory of "sexual selection," were multivalent. Even though Darwin and most other nineteenth-century scientists believed that evolution, like Genesis, demanded women's subservience to men and total devotion to maternity, his theory of evolution contained the seeds of radical interpretations as well as conventional ones. Many feminists and other reformers were keen to these revolutionary insights and embraced evolutionary science as an ally. In fact, a generation of freethinking feminists, socialists, and pioneering sexologists all counted Darwin as an intellectual ancestor in the struggle for women's emancipation, as the following chapters demonstrate.[8] To these social radicals, Darwin's two main contributions were freeing men and women from the legacy of Adam and Eve and redefining the "natural" differences between men and women by placing humans in the category of "animals." Each generation defines what is natural in different terms depending on their scientific and cultural contexts; this book tells how a vocal group of reformist women and men invoked Darwinian evolutionary theory to redefine the natural roles for women in the decades between the Civil War and the outbreak of World War I.

## THE DARWINIAN COSMOS

In 1831, young Charles Darwin set out on his legendary five-year long voyage on the HMS *Beagle*. During his many months at sea and in strange lands, he saw creatures, rock formations, plants, and people that would forever change the way he viewed the world, humans' place in it, and the origins of life on earth. When he relaxed in the evenings or on a slow day, he often read from his favorite book, John Milton's *Paradise Lost*, the epic poem about Adam and Eve's expulsion from the Garden of Eden. Indeed, *Paradise Lost* was the only volume of what might be called "recreational reading" that anyone brought along aboard the *Beagle*.[9] In his book chronicling the *Beagle*'s voyage, Darwin recalled, "Milton's *Paradise Lost* had been my chief favourite, and in my excursions during the voyage of the *Beagle*, when I could take only a single small volume, I always chose Milton." To ensure that he could take *Paradise Lost* with him wherever he went, Darwin even had a special pocket sewn into his coat to hold his pocket-sized edition. Darwin memorized large sections of the poem and sometimes referred to it in describing the many breathtaking things he saw on his voyage, such as the view from the *Beagle* as it approached Buenos Aires: "As far as the eye reached, the crest of every wave was bright; and from the reflected light, the sky just above the horizon was not so utterly dark as the rest of the Heavens.—It was impossible to behold this plain of matter, as it were melted and consuming by heat, without being reminded of Milton's description of . . . Chaos and Anarchy."[10] As Darwin imagined the world of Adam and Eve, little did he know that he would soon introduce an alternative creation story and a brand new way to understand humans' place in the universe.

In *On the Origin of Species by Means of Natural Selection, or the Preservation of Favoured Races in the Struggle for Life* (1859), Darwin cast doubt on the Genesis version of creation, and in *The Descent of Man* he shifted scientific and popular attention away from divine creation and toward a naturalistic explanation of all forms of organic life, including humans. In addition to incorporating humans into the evolutionary saga, *The Descent of Man* also introduced the theory of sexual selection, which Darwin believed explained the differences between men and women, as well as the evolution of heterosexual reproduction. Taken together, Darwin's work helped to usher in a new, evolutionary cosmology based not on special creation and original sin but on individual fitness, reproductive success, and human-animal kinship. Also central to the Darwinian cosmology were the twinned concepts of constant change and unlimited variation—no species was fixed, and the natural world was a wondrous place, subject to all sorts of changes over time. The Darwinian shift to thinking about the world in naturalistic, as opposed to divine, terms had important implications for scientific and popular understandings of gender and sex.

Scientific and cultural debates about evolutionary theory, by definition, invoke larger existential questions: What is the meaning of life? What is humans' place in the universe? What is the natural order of things? Such debates often hinge on, and reflect, ideas about gender. In the nineteenth century, evolutionary theory offered radically new ways to think about the differences between men and women, the function of sexual dimorphism, and the mores governing heterosexual relations (because Darwin focused on the reproductive aspects of sex, his works assumed heteronormativity, although some contemporary readers did see in evolutionary theory the opportunity to make new arguments supporting the "naturalness" of the variety of sexual behaviors found in the animal kingdom and among humans).[11] As a result of evolutionary theory's implications for gender and sex, responses to it were often highly gendered as readers interpreted Darwinian evolution according to whether or not it supported what they believed to be true, or hoped could be true, about men and women.

Popularly accepted views regarding what is natural have particular resonance for questions about gender and women's rights, as the historian of science Londa Schiebinger and others have established.[12] In democratic governments founded on the principle of "natural rights," the political world is supposed to mirror the natural, so what people accept as evidence from nature shapes political, cultural, and personal realities. In the nineteenth century, evolutionary theory did not create new bodies for men and women or offer concrete, unassailable revelations about existing ones, but it did redefine what it meant to be human, and it rephrased questions regarding sexual difference, thereby reframing debates about the rights of men and women. In a Darwinian world, all organisms were not created equally, and it was these subtle differences between individuals that determined who lived long enough to reproduce. Moreover, Darwin described the differences between males and females, especially in *The Descent of Man*, as engines of evolutionary development and hallmarks of advancement. As Darwin explained, sexual dimorphism enabled the evolution of higher animals because it multiplied the possible variations that offspring could inherit, and males were the ones chiefly modified as they participated more fully in the struggle for existence. In his summary of secondary sexual characters in humans, Darwin explained, "We may conclude that the greater size, strength, courage, pugnacity, and even energy of man, in comparison with the same qualities in woman, were acquired during primeval times, and have subsequently been augmented, chiefly through the contests of rival males for the possession of the females." According to Darwin, men's brains had been modified along with their bodies: "The greater intellectual vigour and power of invention in man is probably due to natural selection combined with the inherited effects of habit, for the most able men will have succeeded best in defending and providing for themselves, their wives and offspring."[13] While such statements tended to affirm Victorian ideas about male superiority, keen readers recognized that the shift in base from divine creation to naturalistic evolution and from faith to scientific observation might offer women new possibilities.

The fundamental question, heightened in urgency by Darwin but certainly not introduced by him, was what, if any, lessons about humans could be learned from animals. Thinkers as far back as Aristotle had understood that humans bore obvious structural and other similarities to animals, but Darwin was the first to plausibly propose that neither animals nor humans were specially created by God and that they might in fact be more alike than different. In *The Descent of Man* and his later work, Darwin went so far as to argue that all human traits—including even the ability to reason and express emotions—could be observed among animals and explained by natural and sexual selection, with no help from a divine creator.[14] To Darwin, the central point was not that humans had evolved from animals; it was that humans *were* animals. Later critics have rightly criticized Darwin for imposing Victorian cultural prejudices on the animals and plants he described, but, as George Levine has suggested, perhaps the most accurate word to describe Darwin's philosophy is not anthropomorphism but zoomorphism.[15] By insisting that all human traits could be found to some degree in animals, Darwin stressed the animal nature of humans, not vice versa.

The publication of Darwin's landmark works was not the first time, and surely it will not be the last, that science has provided the impetus for a broad-based rethinking of what it means to be human, male and female. In the seventeenth and eighteenth centuries, the enterprise of modern science itself was constructed along highly gendered lines.[16] Darwin and his contemporaries were, in many ways, the products of this Enlightenment science, and "science" as it came to be practiced and institutionalized in Darwin's era developed along highly gendered and exclusionary lines, as subsequent chapters attest. Darwin was also not the first person to posit the evolution of species, or transmutation as it was then called. His own grandfather, Erasmus Darwin (1731–1802), proposed evolutionary ideas in his canonical and idiosyncratic *Zoonomia* (1794), and the French evolutionist Jean-Baptiste Lamarck's (1744–1829) ideas about the inheritance of acquired characteristics remained popular in Europe and America into the twentieth century. In the United States, Herbert Spencer's "social Darwinism" was equally, if not more, influential than the more naturalistic

Darwinism, especially because it appealed to American's abiding faith in progress and because it did not directly challenge special creation.[17] Spencer aimed to provide one unifying theory, his "synthetic philosophy," for the evolution of everything organic and inorganic—including culture, economic systems, and human societies—that continuously improved toward perfection and was not necessarily based in scientific evidence or observation. Furthermore, among the scientific community at the turn of the twentieth century, as the historian of science Peter Bowler has established, there were actually many competing theories of evolution that rivaled Darwinian natural selection as the most viable explanation of change over time.[18]

Despite the many varieties of evolutionary theory that circulated in the Gilded Age, there was something highly distinctive about the work of Charles Darwin and its American reception, as has been well documented by the legions of Darwin scholars.[19] First, unlike previous scientific attempts to describe the evolution of species, Darwin proposed two viable mechanisms, natural selection and sexual selection, that could have caused such dramatic changes over millions of years. Second, unlike Spencer and the many other popular social evolutionists, Darwin grounded his theories in rigorous scientific observation and experimentation, and they have, by and large, turned out to be accurate. Third, and perhaps most important for the purposes of this study, unlike his predecessors or his contemporary social evolutionists, Darwin fundamentally redefined nature by severing it from an omniscient creator and by placing humans in the animal kingdom.[20] To him, human kinship with animals was so patently obvious that, as he wrote in *The Descent of Man*, "It is only our natural prejudice, and that arrogance which made our forefathers declare that they were descended from demi-gods, which leads us to demur to this conclusion."[21] To date, most historical scholarship on Darwin has focused on the *Origin of Species* and the theory of natural selection. Yet the reception of *The Descent of Man* and the theory of sexual selection are equally rich and deserving of analysis, as a small but growing body of literature has demonstrated.[22] In addition to a focus on women and gender, *From Eve to Evolution* adds much needed attention to the American reception of *The Descent of Man*.

## THE GENDERED RECEPTION OF
## *THE DESCENT OF MAN*

For modern readers to appreciate the gendered significance of Darwinian evolution, imagine what it would be like if scientists today discovered life on another planet that was either sexed differently than humans or not sexed at all. What if extraterrestrial organisms had several sexes, or none at all? What if they reproduced asexually, or homosexually, or both? Might this shed light on current debates about the extent to which biology determines sex and sexuality? In other words, what would happen if what the majority of Americans consider to be the natural order of things—namely, fixed gender and sexual categories—was not preordained or natural after all? Perhaps this was the sort of shock experienced by men and women in the nineteenth century whose interpretation of Darwinian evolutionary theory forced them to reconsider time-honored, biblical prescriptions for male and female behavior, marriage, and reproduction.

In the *Origin of Species* Darwin argued that all species had evolved gradually from a common ancestor, most likely a single-celled hermaphroditic organism, through the process of natural selection whereby those individuals who were the best adapted to their environment would be more likely to survive and pass on their traits to offspring. Darwin only hinted at the ways in which his theory might pertain to humans, famously noting in the concluding pages that one day "light will be thrown on the origin of man and his history," although most of his readers immediately grasped the implications.[23] For one thing, if one accepted Darwin's creation story, there was no such thing as the Garden of Eden, a possibility of particular interest to women's rights activists.

In *The Descent of Man*, Darwin applied evolutionary theory specifically to humans. In response to tremendous pressure from his peers—critics and supporters alike—Darwin's initial goals in writing about human evolution were to explain the divergence of races and the existence of sexual dimorphism. As he continued to compile his notes and thoughts regarding human evolution, however, Darwin realized he would also need to tackle the development of human intellect and morals in order for his completely naturalistic explanation of evolution to be convincing.[24] Such a task proved challenging and often led Darwin to contradict himself as he attempted to construct an evolutionary path, albeit a winding and hesitant one, from protozoa to modern human civilization that explained not only the origin of human life but also its customs and cultural achievements. The mechanism responsible for many of these developments, according to Darwin, was "sexual selection."

Darwin first grappled with sexual selection in unpublished writings from the 1840s, and he alluded to the theory in the *Origin of Species*. There he defined sexual selection not as "a struggle for existence" but as "a struggle between the males for possession of the females," a sort of corollary to natural selection. He claimed that sexual selection accounted for differences in "structure, colour, or ornament" in species where the males and females "have the same general habits of life."[25] But he devoted just two pages to sexual selection. In the years between the publication of the *Origin* and the *Descent*, he continued to puzzle over the persistence of maladaptive traits, traits that conferred no survival advantages to their possessors and, thus, could not be explained by natural selection. Why had traits, such as the peacock's bright plumage, survived?

In *The Descent of Man*, Darwin concluded that maladaptive traits continued to be passed on to future generations simply because the opposite sex found them attractive, thereby increasing the odds that the peacock with the most brilliant plumage, for example, would leave many offspring. The persistence of traits that "must be slightly injurious to the male" convinced Darwin that "the advantages which favoured males . . . leaving numerous progeny, are in the long run greater than those derived from rather more perfect adaptation to their conditions of life." As a result of this revelation, Darwin came to believe that the struggle to reproduce was at least as important, if not more important, than the struggle to survive. This realization seemed to have surprised even Darwin. "It could never have been anticipated," he confessed, "that the power to charm the female has sometimes been more important than the power to conquer other males in battle."[26] Most nineteenth-century naturalists rejected sexual selection theory, but, in the years before his death, Darwin became only more convinced of it. In a letter read before the Zoological Society of London in 1882, just hours before his death, Darwin once again affirmed his belief in sexual selection: "I may perhaps be here permitted to say that, after having carefully weighed, to the best of my ability, the various arguments which have been advanced against the principle of sexual selection, I remain firmly convinced of its truth."[27]

Darwin clarified that sexual selection applied only to instances in which males and females of the same species were exposed to the same conditions and had the same habits, yet one sex, usually the male, had very distinctive traits compared with those of the female to whom he displayed these distinctive traits. Males, for example, often exhibited inordinately brilliant feathers or large tusks, which Darwin reasoned must have appealed to the females, otherwise there would be no adaptive reason for their existence. As one sex (usually the female) repeatedly selected for the desired traits in the other (usually the male), the sexes would differentiate from each other and the desired trait would be passed on to the next generation and exaggerated over time. The two main tenets of sexual selection theory then were male battle and female choice of sexual mates; however, Darwin asserted that among humans, men, not women, selected mates, an observation that puzzled many nineteenth-century reformers because it seemed to contradict Darwin's otherwise firm belief in the animal-human continuum. Darwin's description of mate selection also forced people to examine heterosexual desire in evolutionary and naturalistic terms by suggesting that reproductive choices shaped the evolutionary process, positing links between human desire and animal mating, and proposing that science might help us better understand sexuality and reproduction.

Even though many naturalists remained skeptical of sexual selection theory until the late twentieth century, *The Descent of Man* reverberated widely throughout transatlantic scientific and popular circles. Referring to the theory of sexual selection, the *New York Times* reported, "nothing that Darwin has written is so ingenious or suggestive than the long, minute, and careful investigation in this field."[28] Much to Darwin's surprise, this book did not garner, on either side of the Atlantic, nearly the amount of criticism that had greeted the *Origin of Species*. He mused, "everyone is talking about it without being shocked."[29] Other scientists also noted the equanimity that greeted the *Descent*. Shortly after its publication, Darwin's ally Joseph Hooker informed him, "I dined out three days last week, and at every table heard evolution talked of as an accepted fact, and the descent of man with calmness."[30] One literary notice observed "the very general discussion by the press of Darwin's 'The Descent of Man' has, instead of exhausting public interest in this latest scientific question, greatly stimulated it. The sale of Darwin's work is almost unprecedented in scientific literature."[31] Just a few weeks after the first U.S. editions of the *Descent* hit the stands, Edward L. Youmans, publisher of *Popular Science Monthly*, wrote to Herbert Spencer, "[T]hings are going here furiously. I have never known anything quite like it. Ten thousand *Descent of Man* have been printed, and I guess they are nearly all gone."[32]

Regardless of whether or not readers accepted Darwin's arguments in *The Descent of Man*, all agreed that the book was a literary sensation and a must-read. Even the negative reviews suggested that people read the *Descent*. In its signature ladylike tone, *Godey's Lady's Book*, the popular nineteenth-century women's magazine, noted that the book "will call forth discussion and dissent among the masterminds of the age" but demurred in conclusion, "we are not yet an avowed convert to Darwin's theories, but we find his book exceedingly interesting."[33] The *Galaxy* proclaimed, "[W]hatever may be thought of Mr. Darwin's conclusions as to the origin of man, his book will be found a rich mine of facts, entertaining and curious on the highest questions of natural history."[34] *Old and New* declared the *Descent* to be "as exciting as any novel."[35] *Appleton's* announced that the book was the literary sensation of the month, while *Harper's* observed that "few scientific works have excited more attention" than the *Descent* as evidenced by the fact that one could not open a magazine without reading about it.[36] It appeared on prominent book lists for women's and girls' clubs until the turn of the twentieth century, and the *New York Times* reported that it was among the most popular books checked out of Manhattan public libraries as late as 1895.[37]

While American reviews of the *Descent* often critiqued Darwin's assertion that humans were not specially created by God, several also betrayed a gendered subtext, especially as they tried to make sense of sexual selection. *Overland Monthly* printed the most in-depth analysis of sexual selection in the article "The Darwinian Eden." This review did not so much critique the theory as argue that it could not possibly be a factor in modern society where "the most likely young fellow that ever trod the earth does not stand the ghost of a show beside the rich man, though the latter should be humped as to his back, gnarled and twisted as to his limbs, lean, withered, and decrepit."[38] Other publications took a more circumspect approach to this new theory of sex. *Appleton's* thoroughly explained sexual selection in two consecutive articles but discussed its applications only in relation to birds.[39] "We scarcely know how to deal with Sexual Selection . . . It is both a delicate and a difficult subject, and cannot be discussed within moderate limits," declared the *Albion* before fairly summarizing the theory's main points.

Visual images also presented interesting commentaries on gendered interpretations of *The Descent of Man*. *Harper's Bazaar* published two cartoons in response to the publication of this watershed work. In the cartoon "A Logical Refutation of Mr. Darwin's Theory," a husband read passages from the *Descent* to his wife "whom he adores, but loves to teaze [*sic*]." In the illustration (fig. I.1), the bearded husband kneeled in front of his wife in their well-appointed Victorian parlor and read to her while she cuddled their baby. The wife, however, rejected the assertion that their baby was "descended from a Hairy Quadruped with Pointed Ears and a Tail." "Speak for *yourself*, Jack! *I'm* not descended from anything of the kind," she responded. "I beg to say; and Baby takes after Me. So there!"[40] The accompanying illustration depicted the wife as decorous and civilized, the epitome of nineteenth-century femininity. While bearded, brute man could perhaps have evolved from ape-like progenitors, his refined wife most certainly did not. The second cartoon, "The Descent of Man," played on both racial and

A LOGICAL REFUTATION OF MR. DARWIN'S THEORY.

JACK (*who has been reading passages from the "Descent of Man" to the Wife whom he adores, but loves to tease*). "So you see, Mary, Baby is descended from a Hairy Quadruped with Pointed Ears and a Tail. We *all* are!"
MARY. "Speak for *yourself*, Jack! *I'm* not descended from any thing of the kind, I beg to say; and Baby takes after Me. So there!"

Figure I.1. "A Logical Refutation of Mr. Darwin's Theory," *Harper's Bazaar*,
May 6, 1871, p. 288. Reproduced from the Collection of the
Public Library of Cincinnati and Hamilton County.

gendered anxieties (fig. I.2). The "figurative" man asked the "literal" man why he should care whether or not he was descended from an "Anthropoid Ape," so long as he himself was a man. The literal man, who had simian facial features and who was depicted as speaking in dialect, responded, "Haw I wather disagueeable for your *Guate-Guandmother*, wasn't it?" ["How I rather disagreeable for your great grandmother?"][41] Again, the message was clear: women could not have descended from apes, and no civilized woman would have sanctioned sex with a prehuman ancestor.

Literature, too, provides a window into the gendered reception of *The Descent of Man*. Much turn-of-the-century fiction, notably the work of Kate Chopin, was strongly influenced by *The Descent of Man*, and several other works mentioned the book directly.[42] In her novel *My Wife and I;*

*or Harry Henderson's History* (1871), Harriet Beecher Stowe used sexual selection to grapple with the challenges of courtship and the limited roles for women in the nineteenth century. In an attempt to distract herself from obsessing over Harry Henderson, a love interest, Eva sat down to read her friend Ida's copy of *The Descent of Man*, only to open right to the section on sexual selection, at which point she exclaimed, "Oh horrid!" Far from diverting her from thoughts of Henderson, reading about sexual selection only exacerbated her preoccupation. Ida, her proudly single and

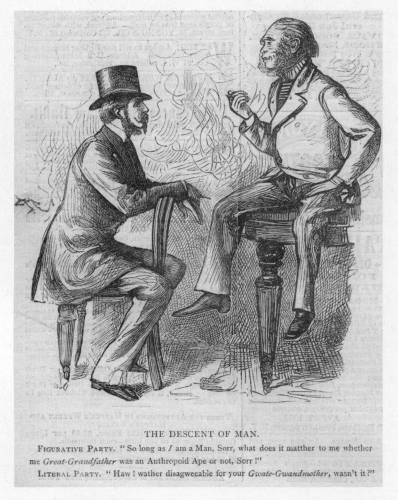

THE DESCENT OF MAN.

FIGURATIVE PARTY. "So long as *I* am a Man, Sorr, what does it matther to me whether me *Great-Grandfather* was an Anthropoid Ape or not, Sorr!"

LITERAL PARTY. "Haw! wather disagweeable for your *Gwate-Gwandmother*, wasn't it?"

Figure I.2. "The Descent of Man," *Harper's Bazaar*, June 28, 1873, p. 416. Reproduced from the Collection of the Public Library of Cincinnati and Hamilton County.

academically oriented friend, encouraged Eva to remain open-minded and read the book for herself, noting that the main reason she could think only of Henderson was that she had nothing else to do.[43] Like the women chronicled in this book, Ida was keen to the new possibilities for gender and sex latent in a progressive interpretation of *The Descent of Man*. In a visual representation of women like Ida, the *Philadelphia Inquirer* ran a cartoon titled "The New Woman Speculating on the Descent of Man," featuring three well-dressed ladies admiring a monkey in a cage, intimating that new women and the acceptance of evolutionary theory went hand-in-hand, perhaps at the expense of traditional male roles.[44] Indeed, for nineteenth-century Americans, the phrase "sexual selection" and the title "The Descent of Man" often functioned as shorthand for new ideas about gender and courtship.

That individuals were keen to the gendered ramifications of evolutionary theory was particularly evident in spoofs parodying *The Descent of Man*. One of the most popular was a song, to the tune of "Greensleeves," first published in *Blackwood's Edinburgh Magazine* and reprinted in numerous U.S. periodicals. Among the "very queer things" that happened as humans descended from animals was that "women plainly had beards and big whiskers at first; While the man supplied milk when the baby was nursed; And some other strong facts I could tell—if I durst—Which nobody can deny."[45] Darwin's suggestion that all organic life had descended from a single-celled hermaphroditic organism troubled some men and women raised on the doctrine of separate spheres and the related idea that, physiologically, women were entirely distinct from men. To others, however, the possibility of a hermaphroditic past sounded exciting and opened up a new world of gendered possibilities.

Perhaps the most colorful response to *The Descent of Man* was the satire entitled *The Fall of Man: Or, the Loves of the Gorillas*, published anonymously by the literary critic and essayist Richard Grant White. Billed as "A popular scientific lecture upon the Darwinian Theory of Development by Sexual Selection, By a Learned Gorilla," this spoof focused on Darwin's assertion that female choice had determined the evolution of species.[46] The "learned gorilla" held a public lecture to explain to his neighbors how their distant cousin had "descended from monkey-hood to humanity." Harkening back to the Genesis creation story, the narrator began by pointing out that, much like humans, monkeys had "fallen" through "the frailty and fickleness of the female sex."[47] In contrast to the biblical account of the fall through female curiosity, gorillas fell through female choice. Once upon a time, the speaker explained, a beautiful female gorilla did not like any of her suitors and refused to be captured. Then one day, she spied a sea serpent, fell instantly in love, and selected him as her mate of choice. Their offspring had tails, and soon tails became a highly desired trait. Subsequently, a whole generation of gorillas with tails evolved. At first gorillas welcomed this development, but the tail kept growing and soon became a tripping hazard. "In this deplorable condition of affairs, we were saved by the action of the same great principle of sexual selection to which we owed our degradation. By a female came our fall, and through a female came our salvation," reported the narrator. Another young gorilla married a tailless hippopotamus and thus reversed the trend for tails. Many generations later, a female resolved to marry a mutant, hairless gorilla, who refused to show interest in any females with hair. Desperate for his attention, she adhered herself to a gum tree for an improvised body wax and, ultimately, gained his affections. In turn, the hairless male gorilla encouraged his other female suitors to remove their hair in the same fashion, and, through these hairless pairings, man evolved from gorilla.[48]

Close analysis of the U.S. reception of *The Descent of Man* reveals the varied ways in which women and men responded to, and in many cases reformulated, Darwin's theory of sexual selection. For those readers who were already inclined to challenge the existing order, Darwin provided the scientific justification to question whether or not patriarchy, monogamy, and female domesticity were in fact natural when so many alternative domestic and sexual arrangements could be found in the animal kingdom. Sexual selection theory also introduced the provocative and potentially radical concept of female choice of sexual partners, providing attentive readers with a new way to think about sexual relations and power systems. Indeed, perhaps the most notable aspect of the American reception of *The Descent of Man* is that so many women enlisted it for feminist purposes.

## WOMEN RESPOND TO DARWIN

Between the 1870s and the 1890s—before the transition to a professional, masculinized science was completed, and before the organized women's rights movement contracted to focus on the vote—an influential group of women spoke and published on the feminist applications of evolutionary theory.[49] Like most nineteenth-century Americans, these women often blended the ideas of Darwin, Spencer, Lamarck, and other evolutionists, often without discerning the differences between them and often referring to all evolutionary ideas as "Darwinian." At the same time, scientists vigorously debated what exactly defined Darwinism, and, today, historians of science continue to wrestle with who, at any given moment, should be considered a Darwinist.[50] In my analysis of the scientific ideas women discussed, I am careful to delineate how these ideas relate to Darwin and attempt to keep the focus on Darwinian ideas, especially sexual selection theory, but I did not preclude from my study sources who wrote about Darwin in ways that were not true to the letter of his word or sources that, for example, blended the ideas of Darwin and Spencer. Doing so would eliminate most nineteenth-century Americans' responses to evolutionary theory.

The women chronicled in this book also tended to be white, middle or upper class, educated, and either unconventionally religious or outright atheists. Although they did not represent a broad swath of American women, they published widely, held powerful posts, and influenced their peers beyond what their numbers might suggest. These Darwinian feminists (my phrase, not theirs) welcomed the entry of science into discussions of women's rights because they thought science provided a better forum than religion to debate sex differences and because they trusted that science could be impartial, even though it often was not.[51] In fact, in their writings, they all compared the Genesis creation story to Darwinian evolution in explaining their preference for evolution. Nineteenth-century Darwinian feminists crafted a compelling case for the feminist applications of evolutionary science and for a feminist approach to biological sex differences, although most of their ideas ultimately fell on deaf ears as women's rights activists shifted to focus exclusively on the vote and as professional science increasingly excluded women. Their writings tell us about the development of "science" as a type of cultural capital and raise important questions about the construction of scientific authority in the nineteenth century. Furthermore, a study of women's enthusiastic responses to Darwin sheds new light on the popularization of evolutionary science in the United States and on the variety of meanings eager readers placed upon this new science.[52]

As readers pondered Darwin's works, the academic departments, institutions, and governmental agencies that today we think of as "science" were all developing. Thus, just as the term "woman" was in flux at the close of the nineteenth century, so, too, was the term "science." Darwin and Darwinian evolution helped shape the development of modern science because his theories popularized the potential of scientific inquiry and inspired public debate about what exactly counted as science, a field that had previously been considered in line with Christian teachings. The women studied in this book did not, for the most part, have access to scientific training or credentials, yet they eagerly read the latest scientific works and believed they were contributing to scientific knowledge. They carved out spaces for themselves to participate in science through women's clubs, which often held discussions of science or sponsored special subgroups on science; through popular magazines that welcomed scientific musings (especially *Popular Science Monthly*); and through the lecture circuit. Their writings document that the exclusion of women from professional science was highly contested and remind us that this historical exclusion continues to have important ramifications for both women and science.

Even though women were, for the most part, excluded from the institutionalization of science, they, too, were inspired by Darwin, especially his materialistic explanation of organic life and his suggestion that humans might be able to learn about themselves from animals. Then, as now, such ideas were difficult to accept even for the most rational, forward-thinking individuals, who were, nevertheless, raised on the twinned concepts of special creation and human distinctiveness. But it was this naturalistic worldview that offered the biggest break from tradition and provided among the most interesting innovations in feminist thought at the turn of the twentieth century. As debates about women's rights increasingly depended on scientific evidence, women frequently used this to their advantage by countering science with what they believed to be better science and by entering the evidence of their own experiences into the scientific record.[53] Evolutionary science was an unlikely and unwitting ally in the struggle for women's rights. Nevertheless, it allowed women to contemplate a world free from gendered biblical restrictions; to ponder sex differences in terms of animals, variety, and change; and to reimagine their bodies and their role in reproduction in an evolutionary, as opposed to biblical, context. From Darwin's example, the women studied in this book also learned to distrust dogma, tradition, and orthodoxy and, instead, view the world around them with a fresh, critical eye and demand verifiable evidence for all supposed truths.

By foregrounding the role that time-honored religious strictures played in motivating the Darwinian feminists and by synthesizing the ways that women interpreted evolutionary science for feminist purposes, *From Eve to Evolution* adds a fresh perspective to existing work on nineteenth-century science and gender, which has tended to focus on the antifeminist uses of science.[54] Feminist historians and philosophers of science have called on historians to recover women's scientific activities in order to help us better understand the construction of science and identify alternative definitions of science. The physicist and historian of science Evelyn

Fox Keller, for example, asserts that the first task of a feminist critique of science is historical: "In the historical effort, feminists can bring a whole new range of sensitivities, leading to an equally new consciousness of the potentialities lying latent in the scientific project."[55] The women studied in this book reveal the latent "potentialities" in sexual selection theory, as well as demonstrate that women, too, were actively engaged in the creation of the American scientific establishment, even as this establishment subsequently excluded them.

The Darwinian feminists also have something important to tell us about the relationship of women to evolutionary science in particular. While for much of the twentieth century many feminists considered evolutionary science to be antithetical to women's advancement, a growing number of scholars now urge a reconsideration of what evolutionary science might mean for women. Feminist theorist Elizabeth Grosz, for example, has encouraged modern feminists to revisit evolutionary theory because "the Darwinian model of sexual selection comes to a strange anticipation of the resonances of sexual difference in the terms of contemporary feminist theory! It provides the outline of a nonessentialist understanding of the (historical) necessity of sexual dimorphism."[56] In *The Nick of Time: Politics, Evolution and the Untimely* (2004), Grosz critiques "the standard, knee-jerk feminist reading of Darwin today . . . [that] he sometimes sounds suspiciously like an apologist for his own culture's masculine privilege," and instead suggests that feminists, and others, look to Darwin's "reconfiguration of culture in light of the fundamental openness he attributes to the natural world." She further proposes that Darwinian evolutionary theory "may be of use to a feminist politics of transformation, which may find his conceptions of time and becoming helpful in rethinking concepts of nature and culture, of human and animal, mind and matter, outside their more conventional feminist frameworks."[57] The nineteenth-century feminists studied in *From Eve to Evolution* were drawn to evolutionary theory because it naturalized a world based on variation and change, established a line of continuity between animals and humans, and probed the boundaries between nature and culture and because they, too, saw in it the potential for nonessentialist, nonreductive accounts of sex difference. Their critical eye toward scientific sexism, however, did not tend to be accompanied by a critical eye toward scientific racism. Like most feminist thought of the time, the Darwinian feminists' ideology was grounded in assumptions of whiteness and, generally, white racial superiority.

Indeed, racial thinking underlies many of the Darwinian feminists' responses to evolutionary theory, as well Darwin's own ideas about gender. In *Darwin's Sacred Cause* (2009), Darwin scholars Adrian Desmond and James Moore persuasively argue that Darwin's strident objection to slavery compelled the publication of *The Descent of Man* and that race was a central concern of the book. Specifically, they contend that Darwin set out to prove, once and for all, that all humans evolved from a common ancestor, a theory called monogenesis. At the time, the more popularly accepted view among scientists and laypeople was polygenesis—the idea that each race sprang from a separate ancestor and should thus be viewed as separate species.[58] Polygenesis was often invoked in defense of slavery, and Darwin found the concept absurd and dangerous. To explain how the various races, often very different in appearance, had evolved from one common stock, Darwin invoked the theory of sexual selection and suggested that each race held a distinct standard of beauty. As men in each race selected mates that best exemplified their specific racial tastes, the races diversified and became more distinct over time.[59] To Darwin and his readers, race, gender, and sex were intimately intertwined from the beginning. While Darwin took great pains to establish the common humanity of all people and protested the most virulent forms of racism of his day, modern readers have rightly noted that, nevertheless, racial hierarchies populate Darwin's evolutionary narrative as humans ascended from "savage" (generally brown) to "civilized" (generally white). The women who were most enthused by Darwinian evolution, as previous scholars have established, also internalized these racial hierarchies and often drew on them to assert that their rightful place was at the top of the evolutionary ladder, together with white men.[60] These racialized assumptions severely limited the radical potential of the Darwinian feminists' critiques of their society, but they are not the only aspect of this story.

Modern scholars debate the extent to which these women's feminist ideals were grounded in assumptions of white superiority and what this means for our interpretation of them today. Among the Darwinian feminists discussed in this book, charges of racism particularly pertain to the writings of Charlotte Perkins Gilman and Elizabeth Cady Stanton. Some scholars, most notably Louise Michele Newman in *White Women's Rights: The Racial Origins of Feminism in the United States* (1999), have argued that evolutionary discourse encouraged latent racist tendencies in the nineteenth-century women's rights movement and that racism is the main legacy of the era's feminist thought. Likewise, literary scholar and cultural critic Alys Eve Weinbaum has suggested that feminists jettison, at least in part, Charlotte Perkins Gilman owing to the racist themes in her work.[61] Other historians, including Judith Allen and Ann D. Gordon, have argued that we should view the Darwinian feminists—in this case Gilman and Stanton, respectively—in historical context and recognize that, for their time, these women were not, in fact, racist compared with the rest of society or even with their reformist peers. They further suggest that even as modern scholars rightly reject the racial undertones of Gilman's and Stanton's work and their assumptions of white racial superiority, we should not dismiss their important contributions to American feminist thought.[62] Moreover, Allen contends that recent critical efforts to dismiss Gilman as "racist" are "not only unhistorical but also antihistorical when its [the Gilman-is-racist school of thought] advocates reject the historian's mission of investigating transformation over time and situating evidence in its own historical context."[63] Taken together, Gilman and the other Darwinian feminists tended to articulate the dominant racial hierarchies of their time and to lobby for reforms that would mainly benefit white women like themselves, sometimes at the expense of people of color. While modern readers recognize the connections between gendered and racial oppression and the ways in which racial ideologies structure gendered ones, and vice versa, it would be ahistorical to discount the contributions of Darwinian feminists because they did not.

Following the examples of Judith Allen, Ann D. Gordon, Michele Mitchell, and others, my work recognizes the racialized thinking among the Darwinian feminists—especially in the places where it is most overt, as in their support of "educated" (read white) suffrage—and seeks to place it in the broader context of their evolutionary and feminist thinking and in the broader historical context in which they wrote. A main contribution of this book is to place religion, science, and gender in conversation with each other, in an attempt to mirror the milieu in which these women interpreted evolutionary theory. As a result, I suggest that their deep frustration with Christian ideology based on Eve, not their internalized racial hierarchies, primarily motivated many women to enlist Darwinian evolution. Furthermore, my research reveals that the Darwinian feminists did not articulate a unified idea of race. The women studied in this book did not all think the same things with regard to race, either as each other or over the course of their long careers. Many of them, especially Antoinette Brown Blackwell, Eliza Burt Gamble, and Margaret Sanger, were particularly concerned with the plight of poor women, who were more likely to not be white, and hoped their visions of reform would especially benefit them.

In addition, the Darwinian feminists pioneered strategies to critique science, reformulate the production of scientific knowledge, and make the scientific enterprise more inclusive, techniques that, ultimately, could be enlisted by other marginalized groups as well. For example, by suggesting that the cultural and the natural were fundamentally intertwined and open to constant variation, Darwinian feminists helped craft the arguments against biological determinism and biological hierarchies, not just for white women but for everyone. Thus, this book suggests that we pay particular attention to the nuanced and complex ways that the Darwinian feminists articulated ideas about sex and race within the broad context of their experiences and their historical realities.

To that end, this study seeks to understand the Darwinian feminists not only in historical context but also in their personal circumstances. Key to such an approach is appreciating the parameters and major events of the women's lives as they lived them. What concerns did they have as women, mothers, and wives? How did these day-to-day, lived realities shape their understanding of evolutionary theory as well as what they hoped evolution might mean for women and men in the future? The women studied in this book had much in common. First, they were all white and middle or upper class, at least in terms of ideology if not always financially, and several were members of the most prominent families in nineteenth-century America. Because of the racism within the women's rights movement, and within mainstream America more generally, African American women did not generally have the opportunity to publish in the women's rights, reform, or scientific periodicals that provide the basis for much of this study, and, subsequently, their voices are largely absent from this work as well. To my knowledge, there are no studies of African American women's responses to evolutionary theory, and I heartily look forward to work in this vein.[64] In addition to shared racial and class perspectives, the women in this book considered themselves working mothers, or at least working wives, which put them in a tiny but growing minority of white, middle-class American women. Educated, high achieving, and ambitious, they each hoped to leave a lasting mark on the world, but they were not sure how, or even if, it would be possible to balance their professional goals with the domestic responsibilities expected of them as women, wives, and mothers. Thus, a common thread that runs throughout their writing—and why they were so intent on looking to animals for alternative domestic models—was the argument that it was natural for women to work outside of the home. They did not agree on how exactly women might do this— have many children, few children, or no children, helpful husbands, no husbands, cooperative domestic arrangements, or professional housekeepers—but they all believed that the progress of women was deeply intertwined with the advance of science and that, in the future, science would enable women to contribute in all realms of life.

## CHAPTER OVERVIEW

Chapter 1 sets the tone for the rest of book by demonstrating how central Eve was to debates about women's rights and why many women eagerly referenced evolutionary theory in general and *The Descent of Man* in particular as an alternative creation story. Ultimately, Darwinian evolution inspired some freethinking (a nineteenth-century term referring to agnostics and atheists) feminists to renounce Eve and Christian orthodoxy all together, forcing a split in the women's rights movement. The women most influenced by Darwinian evolution were, more or less, ousted from the largest suffrage organization, the National American Woman Suffrage Association, partially as a result of their divergent views about the role that religion should play in American culture and within the women's rights movement. After 1890, those women working inside of suffrage organizations wrote less and less about evolutionary science, whereas the feminist women working in freethought, sex reform, and socialist groups continued to publish and speak about the radical potential of Darwinian theory, especially sexual selection. The subsequent chapters chronicle their lives and writings, organized according to the key intellectual themes that the Darwinian feminists advanced.

Chapter 2 analyzes the brief window between 1870 and 1890 when women's rights activists considered science to be a vital part of their agenda and an important tool for their advancement, a development intimately related to the American reception of Darwin. Focusing on the little-known story of Helen Hamilton Gardener's brain donation, this chapter traces the ways in which women used science for feminist purposes and highlights the extent to which discussions of women's rights hinged, quite literally, on the scientific study of women's bodies. This chapter also raises questions about the cultural authority of science, the popularization of science, and the limits of scientific objectivity, questions that were often answered in gendered terms at the turn of the twentieth century.

Chapters 1 and 2 also consider the theme of equality versus difference as articulated by the Darwinian feminists. Were women essentially equal to men, or essentially different from men? Was it possible to be both equal and different? At the heart of the difference question, then as now, was maternity and motherhood. Chapter 3 analyzes how various thinkers applied evolutionary theory to motherhood. Opponents of women's advancement typically claimed that women's foremost function was to bear and raise children; any intellectual or professional endeavors detracted from this sacred duty and imperiled the human race. These arguments were often couched in evolutionary discourse, as exemplified by the much-studied "Race Suicide" panic of the early 1900s. Because of the flexibility of Darwinian discourse, however, evolutionary theory also buttressed a feminist redefinition of motherhood—promoted by Antoinette Brown Blackwell, Charlotte Perkins Gilman, and others—which claimed, in part, that it was unnatural for women to be confined to domestic tasks because female domesticity had no precedent in the animal kingdom. Focusing on feminist applications of animal-human kinship, this chapter examines the turn-of-the-century vogue for fit pregnancy and feminist demands for the reapportionment of domestic duties to enable mothers to work outside the home.

Many feminists and social reformers found that the most provocative idea contained in *The Descent of Man* was Darwin's observation that in all species, except among humans, females selected their sexual mates. To these reformers, restoring "female choice" in humans seemed like a panacea that could solve a variety of social ills, from prostitution, to female subservience, to the excesses of capitalism. Female choice also appealed to reformers because it seemed like a return to a more "natural" state of affairs. Chapter 4 tracks how feminists and socialists utilized female choice to lobby for increased reproductive and economic autonomy for women. Female socialists' trust in science merged with their concerns about the lack of women's reproductive options in the creation of the birth control movement led by Margaret Sanger. Today, as feminists, scientists, and laypeople continue to discuss the relationship between nature and culture, the extent to which biology determines gender, and what a feminist approach to biological difference might be, revisiting the first generation of Darwinian feminists provides both a useful framework and a cautionary tale.

## CHAPTER ONE

# Eve's Curse

And the rib, which the Lord God had taken from man, made he a woman, and brought her unto the man.
—Genesis 2:22

Prior to the introduction of Darwinian evolutionary theory in the late nineteenth century, the Genesis creation story not only revealed the origins of life on earth, it also explained what it meant to be human and, especially, what it meant to be male and female. By the early 1800s, geological discoveries had cast doubt on the literal six days of creation, but, literal or metaphorical, the Garden of Eden still provided the blueprint for the Christian understanding of the universe. While there are in fact two creation stories in the first and second chapters of Genesis, the latter is the one most commonly reiterated. This version explains that Eve was made from Adam's rib to be his "helpmeet."[1] Soon thereafter Eve caused the couple's exile from the Garden of Eden by disobeying God's word, eating fruit from the tree of knowledge, and successfully encouraging Adam to follow suit. As punishment, God sentenced Adam to a life of toil in the land outside of Eden. To Eve, God thundered, "I will greatly multiply thy sorrow and thy conception; in sorrow thou shalt bring forth children; and thy desire shall be to thy husband, and he shall rule over thee."[2] For generations, the legacy of Eve's secondary creation, sin, and subsequent curse shaped church doctrine, public life, and popular culture, informing individual's images of themselves and their ideas about what was possible for women and men. Thus, to fully understand women's responses to evolutionary theory, we must begin with Eve. Indeed, the most fundamental and perhaps most urgent reason why many women drew inspiration from evolutionary theory, at least initially, was that it provided an alternative

creation story to the Garden of Eden, although, by 1890, the perception
of evolution as an alternative to Christianity forced a split within the
women's rights movement.

## THE LONG LEGACY OF EVE

Nineteenth-century Americans could expect to hear about Adam and Eve
in church, read about them in popular periodicals and literature, and see
them depicted in art.[3] In 1833, two paintings entitled "Adam and Eve"
and "Paradise Lost" (billed together as "The Temptation and the Expulsion of Adam and Eve") by the French artist Claude-Marie Dubufe toured
the United States and "probably drew together greater crowds of specta-
tors than any pictures ever exhibited in the country" (figs. 1.1 and 1.2).[4]
In New York City alone, twenty-five thousand people were said to have
paid admission to view the paintings.[5] One reviewer concluded from the
"vast number of visitors, old and young, of both sexes, that throng by day
and night" to see the pictures on display in Philadelphia that "we may be
warranted in supposing that the work of no one artist ever before afforded
American taste such perfect gratification." This same reviewer observed
that Adam and Eve "are interesting in the highest degree to all the human
family" because they revealed "the facility of a Temptation to which all
the sons of earth fell victims through their beautiful mother, and the ag-
ony of an Expulsion, in the endurance of which the intellectual energy of
the world's Father sustained and comforted the winning woman for whom
he sinned and suffered."[6]

Decades later, William Dean Howells, the legendary nineteenth-
century writer and influential editor of the *Atlantic Monthly*, attributed
his lifelong interest in art to having seen these very paintings.[7] At the turn
of the twentieth century, Howells' friend and colleague Mark Twain published two volumes presenting, in a modern, humorous way, the diaries of
Adam and Eve.[8] In the intervening seventy years, it had become culturally
acceptable for Twain to satirize the biblical pair, but the key to Twain's
humor was that most people were still deeply invested in this ancestral
relationship. Twain also proposed, tongue-in-cheek, that the town of El-
mira, New York, erect a monument to Adam, since in "tracing the genesis
of the human race back to its sources [in *The Descent of Man*], Mr. Darwin had left Adam out altogether."[9] Even as artists like the irreverent
Twain toyed with the Garden of Eden story, the original couple informed
the stories and images people conjured when contemplating women's role
in society.

Figure 1.1. Claude-Marie Dubufe, "Adam and Eve" (1827). Courtesy of the
Musée des Beaux-Arts, Nantes, France. Photo credit: Art Digital Studio.

In private life, too, Adam and Eve shaped Americans' ideas about what
it meant to be man and woman. The historian Anthony Rotundo found
numerous references to the biblical pair in his research on conceptions of
manhood in the nineteenth century, especially in letters and memoirs (in
addition to more public sources). The Bible was the most frequently read
book in nineteenth-century America, yet we do not often think of it as

Figure 1.2 Claude-Marie Dubufe, "Paradise Lost" (1827). Courtesy of the
Musée des Beaux-Arts, Nantes, France. Photo credit: Art Digital Studio.

a marriage guide. Rotundo's research demonstrates that, in fact, biblical
passages on Eve influenced the parameters of many couples' relationships.
Not surprisingly, Rotundo found letters written by men "invoking the
Bible . . . to support the husband's power" over the wife.[10] He also located
several references to Eve in letters and memoirs written by men. Shedding
light on the role that the biblical creation story played in shaping male at-

titudes toward women, most of these references to Eve described women as "temptresses."[11] To nineteenth-century readers, the most important message about marriage to be gleaned from the Bible was that God intended for the husband to be the head of the household and, by extension, the nation. As Rotundo observes, "[B]efore a woman defied her husband or dealt with him on equal terms, she had to struggle with the force of biblical injunction and with the centuries of marital tradition that were justified by those injunctions."[12] A daunting proposition indeed.

The narrative and imagery of Adam and Eve was so deeply ingrained in American and European culture that Eve played the pivotal role in debates about women's rights from the seventeenth century, when women began to publicly demand more opportunities, to the twentieth, when they focused on and secured the right to vote. References to Eve reached a fever pitch in the nineteenth century during periods of heightened publicity or success of the women's rights movement: in the 1840s and 1850s, and then again in the 1880s and 1890s. Regardless of the particular question at hand, women were told they were not fit for public or professional life and that they must remain subordinate to men as a result of Eve's secondary creation, transgression, and curse. To be sure, women who agitated for increased educational, personal, and professional opportunities encountered many obstacles, but the one seemingly impenetrable barrier that generation after generation had to confront was the legacy of Eve. Even antifeminist arguments that did not explicitly mention Eve were grounded in the basic premise that women were created as an afterthought and destined for treachery. Opponents of women's rights often drew on the New Testament writings of Paul, for example, but these passages were informed by Eve's conduct in Eden and generally served to remind audiences to heed the lessons in Genesis. As many women's rights advocates noted, Eve provided the foundation from which all other ideas about women developed.

In the decades leading up to the Civil War, the women who dared test the boundaries of their limited sphere, the relatively few that there were, faced their most vocal opposition from members of the clergy, men who were well suited to argue the Bible against women's rights. When pioneering abolitionists and women's rights activists Sarah and Angelina Grimké first spoke in public in the late 1830s, clergymen banded together to bar them from churches and mobilize public opinion against them. In 1837 the Massachusetts Congregational clergy issued a public letter warning that when "a woman assumes the place and tone of man as a public reformer . . . her character becomes unnatural."[13] Nearly twenty years later, at the Fifth National Convention for women's rights in 1854, the activists

cited continued clerical opposition as a singular hindrance to the movement, resolving unanimously, "[W]e feel it a duty to declare in regard to
the sacred cause which has brought us together, that the most determined
opposition it encounters is from the clergy generally, whose teachings of
the Bible are intensely inimical to the equality of woman with man."[14] To
men of the cloth, and indeed to the vast majority of Americans, women
speaking in public or, worse, on behalf of their own rights violated the
most essential facts of God's divine order, the very same order that provided the blueprint for democratic government and public affairs.

Since the Enlightenment, debates about the ideal political order have
drawn inspiration and justification from what was seen to be the divine,
natural order in the Garden of Eden. As the historian Nancy Isenberg establishes in *Sex and Citizenship in Antebellum America* (1998), "the
creation story and the state of nature played a continuing, vital part in
antebellum political discourse."[15] At the Virginia Constitutional Convention of 1829, to give just one example, delegate Abel Upshur argued
against equal suffrage for men and for the timelessness of a "feeling of
property," which, naturally, made some men the rulers of others, as proof:
"Adam was the first of created beings; Eve was created next; and the very
fiat which brought her into existence, subjected her to the dominion of her
husband. Here then was no equality."[16] In depictions of the ideal political state, Eve's secondary status and propensity to sin provided irrefutable
evidence against women's rights, including but not limited to voting. In
1849 the abolitionist Richard Henry Dana lectured on "Woman" in Philadelphia. According to coverage of his speech in the women's press, Dana's
main point was that women could only "stand in awe and reverence of
man" because Adam was the "first man," forever sealing women's fate as
secondary and ancillary creatures.[17] Even though he defended women's
right to petition, the antebellum statesman, abolitionist, and sixth president of the United States John Quincy Adams denied women rights as
equal citizens; such a proposition simply went against God's creation. In
his 1842 lecture, *The Social Compact*, Adams explained that in order to
understand the ideal plan for democratic government, one needed to look
no further than the Garden of Eden. According to Isenberg's analysis, Adams reasoned that "Adam and Eve introduced civil society into the state of
nature, and that their union symbolized the universal model of bourgeois
society." Eve brought conflict into the Garden of Eden, explaining why
men and women should not both be involved in politics. Adam, on the
other hand, served as a "cautionary tale about allowing women too much
political influence."[18]

Debating God's plan for the universe was a tall order. Nevertheless, pioneering feminists, from Judith Sargent Murray (1751–1820) to Sarah Grimké (1792–1873) to Elizabeth Cady Stanton (1815–1902), all challenged the "rib" story. Antebellum feminists reinterpreted or dismissed Eve in their writings as a way to stake a claim for women's increased participation in public and private life, but, in an era when women could not hold leadership positions in church or state, this tactic met with limited success. According to Murray, one of the first American authors to write on behalf of women, men, rendered "blind" by "self love," were too "wholly absorbed in a partial admiration of [their] own abilities" to notice the real moral in the Garden of Eden tale: Eve ate of the apple to gain knowledge, whereas Adam did so simply because Eve invited him to. "Thus it should seem," Murray concluded, "that all the arts of the grand deceiver . . . were requisite to mislead our general mother, while the father of mankind forfeited his own, and relinquished the happiness of posterity, merely in compliance with the blandishments of a female."[19] Seen in this light, Eve was intellectually curious while Adam was a fool. Sarah Grimké, the antebellum abolitionist who insisted on women's right to speak in public, believed that Adam and Eve bore equal responsibility for their fall from grace and, thus, that they were intellectual equals as well. According to Grimké, "the welfare of the world will be materially advanced by every new discovery we make of the designs of Jehovah in the creation of woman."[20] Twenty years after she wrote, evolutionary theory became one such "new discovery."

Before they could draw on Darwinian evolutionary theory, women countered antifeminist invocations of Eve by citing the first chapter of Genesis, which describes men and women as simultaneous creations. Lucretia Mott, for example, quoted these verses in replying to Richard Henry Dana's 1849 remarks on women as related to Eve. Women claimed that their simultaneous creation made them "co-equal" with men, a powerful intellectual and rhetorical move. Throughout the 1850s, women continued to cite simultaneous creation, coequality, and cosovereignty to justify their campaigns for political inclusion. According to Isenberg's comprehensive study of antebellum feminist thought, coequality must be understood as a "conceptual revolution." These early feminists rewrote the social contract and "carved a theoretical space for women within the imaginary script of the 'original contract' in the state of nature" because their notion of "simultaneous creation challenged the gender asymmetry that enlightened thinkers had firmly rooted in the state of nature."[21] By the 1870s, feminist arguments for simultaneous creation and coequality, along with those questioning the relevance of Eve more generally, enjoyed the support of the new science of evolution, but arguments linking women's degraded status to Eve persisted.

The mainstream consensus that women's lot in life was forever fixed by Eve's transgression survived the vast cultural upheaval of the Civil War seemingly unscathed and offers perhaps one reason why demands for "universal suffrage" for African Americans and women met with little success during and after the war. In 1873, the Transcendentalist-turned-Catholic Orestes Brownson concluded that women were not fit to rule themselves, let alone others, because "Revelation asserts, and universal experience proves that the man is the head of the woman, and that the woman is for the man, not the man for the woman; and his greatest error, as well as the primal curse of society is that he abdicates his headship, and allows himself to be governed, we might almost say, deprived of his reason, by woman."[22] As another opponent of women's rights succinctly explained in 1869, women were prima facie inferior to men because: "1. Her creation was subsequent to that of man. 2. The first woman was taken from the side of man. 3. Her creation was avowedly to supply man with a companion. 4. She was of the sex which implies maternity."[23] Case closed.

In 1871, the feminist paper *Woodhull and Claflin's Weekly* lamented women's limited options when it came to confronting arguments based on Eve. "There is one argument urged in favor of man's right to rule in the political world, and against women's right to participate in the business of legislation, that has never been fully met . . . by the advocates of woman's enfranchisement" observed the author. "The doctrine of the so-called 'Fall of Man,'" the article continued, "has always been the most effective weapon the believers in the divine authenticity of the Scriptures have wielded against the recognition of her equality. Indeed, it is the only basis of nearly all they have to say on the subject."[24] Almost in direct response, the editors of *Godey's Lady's Book*, the most popular women's magazine of the century, criticized "[t]he efforts of that small band of women who assume to represent their sex in claiming the right of suffrage." These women "have so persistently ignored the great and radical differences between the sexes that it is especially necessary to recall them." To understand these differences, women needed only consult "the doctrine of the Bible," which explained "that when banished from Eden, *man* was ordained to be the worker, inventor, and maker of things from earth; the provider and protector for the household; the lawgiver and defender of social, moral, and political rights, the sustainer of moral and religious duties." Women, on the other hand, "reign[ed] supreme" in "the Kingdom of

Home" as "the preserver of life, the first teacher of manhood, the guardian of home, honor, and happiness."[25] What could better support arguments against women's increased participation in public life than the sense that God Almighty had created woman from man's rib to be his helper only for her to defy His instructions and cause the downfall of humankind?

## THE RESURGENCE OF EVE IN THE 1880s AND 1890s

To many people experiencing the fast-paced cultural and industrial changes that characterized the late nineteenth century, women's rights and evolutionary theory were intimately connected as modern developments. To be sure, women's rights activists were not necessarily evolutionists, and the vast majority of male evolutionists were certainly not feminists. Yet, to many observers, feminism and Darwinism were bound together as examples of new ideas that threatened to disrupt the traditional order. Feminism and Darwinism also shared a crucial link in that both necessitated a reevaluation of the Genesis creation story. For women to gain more rights and opportunities, old associations with Eve needed to be cast away; likewise, for those who took Darwinian evolution seriously, a reconsideration of the literal Garden of Eden was also in order. The connection between women's rights and evolutionary theory was often invoked by women's rights activists, as well as by those who opposed both women's rights and evolution. In a pamphlet titled "Woman's Rights" (1867), the Reverend John Todd (whose byline boasted that he was also the author of the aptly titled "Serpents in the Doves' Nest") traced the connection between women's rights and evolutionary theory, noting that both epitomized his generation's "tendency to break up old associations" and their desire "to be emancipated from the beliefs of our fathers." Men of his generation, Todd charged, "would rather feel relieved to have you convince them that they sprang from a race of apes and gorillas." Among women, on the other hand, "there is a wide-spread uneasiness,—a discontentment with woman's lot, impatient of its burdens, rebellious against its sufferings, an undefined hope of emancipation from the originary lot of humanity by some great revolution, so that her condition will be entirely changed!" But, of course, woman could never be "independent and self-supporting" because "God never designed she should be." "Any other theory is rebellion against God's law of the sexes, against marriage, which it assails in its fundamental principles, and against the family organization, the holiest thing that is left from Eden," thundered Todd.[26] Writing in 1867, Todd was the harbinger of a much larger effort by cultural conservatives to defend the literal Genesis creation story against challenges from religious moderates, women's rights activists, and, now, evolutionists.

Prior to 1875, as the historian of religion Jon H. Roberts has documented, Protestant opponents of evolutionary theory assumed that Darwin's ideas about transmutation would be dismissed by scientists, much like all previous theories of evolution had been before.[27] As a result, they did not spend too much time discrediting Darwin's work or preaching about its negative implications. They trusted that scientists, heretofore their allies, would do this for them. When it became clear by the mid-1870s that Darwin's work was different from previous theories of organic evolution and that the majority of scientists had, in fact, accepted it, orthodox Protestant thinkers realized they had a problem on their hands. To counter the growing scientific consensus in favor of Darwinian evolution, Protestant opponents focused not on the theory's scientific shortcomings—this, they realized, was beyond their expertise—but rather on drawing a distinct line between religion and science: either one was on the side of God and the Bible or one was on the side of Darwin and atheism. To evangelical opponents of Darwin, there was no longer a middle ground. And they undertook a vast public relations campaign to convince Americans that they, too, had to choose between God and Darwin, a strategy that had important implications for the women's rights movement as well as for the American reception of evolution.

While most Protestant intellectuals managed to accommodate their Christian beliefs with evolutionary theory, a vocal minority of opponents honed in on evolution's challenge to special creation and waged war against Darwin. To these Protestant thinkers, God's purposeful creation of humans was the glue holding together the entire Christian belief system. Not only did God's creation of Adam and Eve demonstrate his personal involvement in the world, it also proved that human beings were made in God's image. This was a crucial point. Unlike twenty-first-century Creationists, nineteenth-century antievolutionists did not stress the literal six, twenty-four-hour days of creation; to them, the important thing was that God had personally intervened in the world to create human beings in his likeness. Furthermore, these ministers argued, if Adam and Eve did not fall from grace, then the rest of the Bible, including redemption through Christ, was for naught. Even though Protestant intellectuals disagreed about the exact meaning of the Garden of Eden, "many of them," as Roberts argues, "could agree that the scriptural account of early human history was the linchpin of a proper understanding of the introduction of sin in the world, its transmission from one generation to another, and the need for the divine grace they believed was incarnate in Jesus."[28] To this line of thought, abandoning a belief in special creation meant impugning the sanctity of the entire Bible. Moreover, included in a belief in special creation were the related convictions that all species were fixed in perpetuity (because God had made each perfect in its own way) and that a tremendous gulf—moral, intellectual, emotional—separated humans from animals. Darwinian evolution challenged all three of these linked beliefs.[29]

To highlight the threat Darwin posed to believers and wayward believers, Protestant opponents delivered countless sermons and published numerous pamphlets extolling the "argument from design" and the Genesis account of creation. As Roberts explains, "In the judgment of many Protestants, challenges that the transmutation hypothesis posed to the veracity of the biblical narrative constituted its most dangerous and alarming feature," especially with regard to the "origin of humanity."[30] Stressing special creation as the foundational building block of Christian faith allowed opponents of Darwin to argue that evolutionary theory was inherently atheistic, no matter what the namby-pamby moderates would have one believe.

The evangelical campaign against Darwinian evolutionary theory was part and parcel of a larger movement at the end of the nineteenth century to define America as a "Christian" nation. Protestant reformers and political leaders also lobbied Congress to declare Christianity the official religion of the United States in a Constitutional Amendment that stated, "Almighty God as the source of all authority and power in civil government, the Lord Jesus Christ as the Ruler among nations, and His Will, revealed in Holy Scriptures, as of supreme authority." In addition to the failed effort to make Christianity the official religion of the United States, reformers also attempted, with varying degrees of success, to strengthen censorship provisions, enforce Sunday closing laws, and teach Protestantism in schools. In her comprehensive study of nineteenth-century censorship law, Helen Lefkowitz Horowitz describes the second half of the nineteenth century as marked by "intense efforts to define the nation as Christian," partially in response to the perceived threat to orthodoxy posed by Darwin.[31] Taken together, these evangelical and reform efforts inspired broad cultural conversations about the significance of the Genesis creation story to the American way of life.

Throughout the 1880s and 1890s, in response to the twinned threats of Darwin and feminism, religious leaders once again called upon Eve to set the record straight with regard to creation and women's appropriate place in society. Evangelicals in particular responded to the challenges of modernity with calls for a "muscular Christianity," which was also highly gendered.[32] Emphasizing woman's creation as an afterthought and her sinful behavior in the Garden of Eden, nineteenth-century ministers "championed the 'rib' story," in the words of historian Kathi Kern, to settle the woman question, as well as simultaneously bolster biblical adherents whose faith might have waned as a result of Darwin's publications.[33] According to Kern's research, countless biblical commentaries published in the 1880s and 1890s emphasized the manliness of Adam and the femininity of Eve as the exemplars for modern life. Biblical scholars repeatedly cited Eve's curse to suffer in childbirth and be subordinate to her husband as the final word on woman's secondary status. As one biblical scholar contended, "all subsequent passages of the same import are but repetitions and expansions" of this one.[34] Evangelicals described Eve as inherently weak, sinful, animalistic, and "naturally subordinate," traits they also ascribed to modern women.[35]

Evangelical Protestants were joined in their opposition to women's rights by a wide array of political, business, and reform leaders whose own lives were changing in unprecedented ways. The final decades of the nineteenth century witnessed cyclical economic uncertainty, record numbers of labor strikes, and general political instability, forces that coalesced into what some historians have described as a "crisis in masculinity."[36] Furthermore, with what appeared to some to be stronger, more virile men arriving in the United States from Eastern and Southern Europe, native born, middle-class, white men—those who were increasingly confined to unmanly desk jobs—felt doubly threatened. The loss of professional autonomy, coupled with the decrease of physical labor and the influx of immigrants, caused some men to fear for their own virility and status in the political and social hierarchy. Leading public figures including the future president Theodore Roosevelt, himself a recovered "weakling," urged men to head out West and recapture their manhood by participating in manly pursuits like wilderness exploration and hunting.[37] With their status ostensibly in jeopardy, many middle- and upper-class white men were particularly threatened by what they saw as the encroachment of women on the previously male-only spheres of higher education, the professions, and political life.

To defend against these perceived threats to white masculinity, business and political leaders joined evangelicals in invoking Eve to remind women of their sacred, timeless duties. As former President Grover Cleveland wrote to the *Ladies' Home Journal* in 1905, "Those who . . . [seek] to protect the old and natural order of things as they relate to women reverently appeal to the division of Divine purpose clearly shown when Adam was put in the Garden of Eden to dress it and keep it." Readers should remember, too, that "Eve was given to him as a helpmeet and because it was not good that man should be alone." As further enticement, Cleveland encouraged Americans to remember the curse cast upon Adam and Eve for disobeying this divine order.[38] During such confusing times, President Cleveland spoke for the legions of white men who sought comfort and order in the patriarchal gender roles outlined in the second chapter of Genesis. Opponents of women's rights and antievolutionists, often one and the same people, were deeply invested in preserving popular faith in the Genesis creation story and resisting the aspects of modern life that threatened to upend it. In an important sense then, what linked women's rights and evolution together in the public imagination was Eve.

## EVOLUTIONARY THEORY OFFERS AN ALTERNATIVE TO ADAM AND EVE

After generations of being told that the Bible provided the ultimate, immutable justification for female subordination, late nineteenth-century female activists welcomed new, scientific gender paradigms that did not focus on who said what to whom in the Garden of Eden. In 1875, the *Woman's Journal*, the official paper of the American Woman Suffrage Association (AWSA), ran an article triumphantly proclaiming that evolutionary theory heralded a "new day" for women. The author, Claire, enthusiastically reported that if one accepted Charles Darwin's ideas, "Woman can no longer be taunted with having brought on humanity the traditional curse." "Is not the idea fraught with the possible promise of a new day for womankind?" she exclaimed. Women would not be able to learn, work, or vote on an equal basis until "the time-worn views concerning Woman's connection with the fall of Man, and hence with all of human suffering and sin shall cease to be entertained." Evolution promised to excise these "time-worn views." Claire lamented that most scientific men ignored, at best, the feminist implications of evolutionary theory, but she concluded "with a sublime faith in the future, that one Utopia of human dreams, we lay aside our doubts and fears and perplexities, and rest in the shadow of that rock of reason—the 'survival of the fittest.'"[39] To Claire, debating women's rights in terms of reason and natural selection, rather than the Garden of

Eden, boded well for women's advancement. As the nineteenth century drew to a close, Claire was just one of many women celebrating the introduction of evolutionary theory into debates about women's rights.

Women's club and women's rights networks probed the nature of sex differences and debated whether answers to the vexing "woman question" could be found through religion, science, or both. Many noted that change was afoot as questions previously answered by the Bible could now also be debated in the realm of evolutionary science. Prominent women's rights activist and abolitionist Thomas Wentworth Higginson compared religion and science to two chivalrous knights dueling to determine which was better able to take care of, and define, women. Even though he saw grounds for hope in both religion and science, ultimately he threw in his lot with science, even as he remained skeptical of male scientists.[40]

Freethinking feminist Helen Hamilton Gardener also linked the cause of women's rights to evolutionary theory in her 1885 essay, "Men, Women and Gods." Like most of her peers, she believed that women's degraded position was related to ideas about Eve. With her characteristic frankness, she observed, "It is always a surprise to me that women will sit, year after year, and be told that, because of a story as silly and childish as it is unjust, she is responsible for all the ills of life." "That because, forsooth, some thousands of years ago a woman was so horribly wicked as to eat an apple," Gardener continued, "she must and should occupy a humble and penitent position, and remain forever subject to the dictates of ecclesiastical pretenders." Luckily, however, "The morals of the nineteenth century have outgrown the Bible. . . . What Moses and David and Samuel taught as the word and will of God, we, who are fortunate enough to live in the same age with Charles Darwin, know to be the expression of a low social condition untempered by the light of science."[41]

Most women's rights activists were not as freethinking as Gardener, and they tended to use evolutionary principles as a way to interpret, not reject, the Bible. Often women cited evolution as evidence for the first chapter of Genesis (simultaneous creation) or as a way to argue that focusing on Eve's transgression denied women the many other important, especially reproductive, roles they played in life. These women attempted to blend science with religion to better understand human creation and sex differences. At an 1869 women's rights convention held in Newport, Rhode Island, Isabella Beecher Hooker, an active suffragist and the sister of Harriet and Catharine Beecher, delivered a speech about the relation of the Bible to women's suffrage. After arguing that Genesis, properly interpreted, was a story of gender equality, she attempted to take her argument "a step further than this, and presented a number of scientific facts to prove that the highest types of vitality take the female form."[42] By this time, the Bible could no longer stand on its own as the definitive source of information about gender or human origins; instead, Hooker buttressed religious doctrine with examples from nature and "scientific facts."

Similarly, Illinois lawyer and suffrage advocate Catherine Waugh McCulloch applied evolutionary principles to Genesis. In a pamphlet entitled "The Bible on Women Voting," McCulloch reasoned, "[T]he scientists of today quite agree with the Genesis parable concerning the creation; that creation was in the ascending scale, first the lower creatures, then the higher animals, then man, and last at the apex the more complex woman." Read in this light, the order of creation did not support female subordination, but, "it might rather be a reason why men should obey women."[43] In combining the biblical and evolutionary accounts of human origins, McCulloch upended the antifeminist tactic of dismissing women's demands by linking them to Eve's secondary creation. To the contrary, she argued that, from an evolutionary standpoint, Eve's creation from Adam provided evidence for female superiority.

Emily Oliver Gibbes echoed this sentiment in *The Origin of Sin and Dotted Words in the Hebrew Bible* (1893). Here she took Paul to task for interpreting Genesis literally and taking pride "in the fact that Adam was first formed, then Eve." "In these days of belief in evolution it is the other way," Gibbes proclaimed. "If Eve evolved from Adam, she was higher than man" in the order of organic beings.[44] In her compilation of progressive biblical commentaries written by ministers, Frances Willard, the powerful president of the Woman's Christian Temperance Union (WCTU), included one that also blended the evolution and Genesis accounts of creation to argue for women's superiority: "if we find God gradually advancing in his work from the inorganic earth to the mineral kingdom, then to the vegetable kingdom, and last of all making man, the fact that woman is made after man suggests her higher qualities rather than man's superiority."[45] Other advocates of women's rights interpreted evolution to mean that Eve never existed. As Frederic Hinckley, minister of the Free Religious Society of Providence, Rhode Island, wrote in his pamphlet, "Woman Suffrage in the Light of Evolution" (1884), "Eve was not made from one of Adam's ribs, but both have been evolved out of that Universal whose mysteries we cannot fathom, but which we may be sure knows no subjection of the one to the other, having made of one blood all classes and conditions of men."[46]

Throughout the 1870s and 1880s, women's rights advocates cheered Charles Darwin for exposing, once and for all, the fraud of the "rib story."

They believed that evolutionary theory marked a turning point in discussions of gender, one that would be favorable to their cause. For centuries, men and women had debated the meaning of Eve, and, even at the dawn of the twentieth century, the lessons drawn from the Garden of Eden still circumscribed women's opportunities. By presenting an alternative creation story, evolutionary theory offered the potential to revolutionize popular thinking about gender and sex difference.

## RACE-BASED EVOLUTIONARY HIERARCHIES, EDUCATED SUFFRAGE, AND THE TURN TO DIFFERENCE

By the early 1890s, however, changes within the women's rights movement and within the broader American culture made it less acceptable for women's rights advocates to openly espouse Darwinian evolution or apply science to questions of sex difference. For one thing, opponents and supporters of evolutionary theory hardened their positions, and evolutionists themselves divided into several, often competing, schools of thought— making it more confusing for women to advocate one solid evolutionary position.[47] Furthermore, in the decades following the Civil War, rapid industrialization, corporate consolidation, and economic uncertainty characterized the U.S. economy as well as U.S. culture, in what one historian has famously described as "the incorporation of America."[48] Such vast cultural changes institutionalized and masculinized the definition of science, as chapter 2 will discuss, as well as permanently altered the terrain of the women's rights movement.

Perhaps most important for the purposes of this study, between 1875 and 1890, the women's rights movement transformed from a splintered, fringe element in American culture into a powerful voice in American public life, one that enjoyed the support of thousands of women working together under the auspices of the reunited National American Woman Suffrage Association (NAWSA).[49] Prior to 1890, two competing organizations represented the women's rights movement: the American Woman Suffrage Association (AWSA), which advocated that suffrage be won state by state, and the National Woman Suffrage Association (NWSA), which advocated broad-based feminist reforms at the federal level. Led by the iconoclastic Elizabeth Cady Stanton, NWSA also critiqued marriage and the church, much to the chagrin of the AWSA leaders. The 1890 merger of NWSA and AWSA represented the triumph of the less heretical AWSA vision of women's rights, which focused mainly on securing the vote.

Whereas critiques of the Bible, the clergy, and orthodox Christianity had been a foundational element of antebellum feminist thought, by the 1880s such arguments were frowned upon and ultimately jettisoned from the formal women's rights arsenal as leaders prioritized mainstream appeal and expediency over radical critiques of patriarchy. In important and understudied ways, evolutionary theory was one factor in this tactical and rhetorical realignment of the women's rights movement.

The more measured tone of the post-1890 women's rights movement was amplified by the growing influence of the WCTU and its long-serving president Frances Willard. By far the largest and most powerful women's group of the era, the WCTU eventually came to regard suffrage for women as vital to its larger goal of "home protection." Specifically, they believed that alcohol and other vices would be banned a lot sooner if women had a voice and a vote in the legislative process. For their part, NAWSA leaders relished the possibility of joining forces with the WCTU, an organization that counted over two hundred thousand members compared with the suffragists' ten thousand.[50] As a result of this new type of member, NAWSA leaders shifted gears to agitate for suffrage within the bounds of mainstream, Christian values. At the same time, the WCTU brought a new evangelical emphasis to women's rights rhetoric, as Kathi Kern describes. WCTU materials bore titles such as "Jesus the Emancipator of Women" and emphasized that it was Christian women's duty to vote, largely in order to outlaw alcohol. For Willard and the large influx of WCTU members of NAWSA, securing the vote for women was part of God's divine plan for Christian women to help purify politics and society.[51] To these women, it did not make strategic or logical sense to dispense with the Bible or accept a materialist science that taught that humans were part and parcel of the animal kingdom, leaving Stanton and her freethinking colleagues out of step with NAWSA.

As it became less and less acceptable to openly critique Christian doctrine on behalf of women's rights, women's uses of evolutionary theory shifted as well. Women's initial enthusiasm for Darwin had emerged in various forms—from wholesale adoption to blending with Christianity—but, by the late 1880s, Darwinian arguments for women's emancipation were most often advanced by women working outside of, or on the periphery of, NAWSA. NAWSA members, on the other hand, increasingly drew on theories of social evolution—those written and inspired by Herbert Spencer and his protégés including the American William Graham Sumner—and less on the nonteleological, naturalistic evolution of Darwin. The women themselves, however, did not often distinguish between Darwin and Spencer and tended to blend both schools of thought together, especially in support of their belief that evolution meant progress. As the historian Jackson Lears explains, "Much American thought in the early twentieth century combined the delusion that Darwinian theory underwrote linear human advance with a vague technological determinism. From this implicitly reformist view, social values as well as political and economic institutions had simply not 'evolved' far enough to keep up with the realities of human experience."[52] Social evolutionary arguments tended to sever evolutionary theory from its radical, materialist implications and in many cases from experimental science itself. Instead, adherents promoted a progressive, goal-oriented evolution that could be merged more easily with Christianity as a tale of triumph for "civilized" (which often meant white) people. Arguments grounded in Spencer's work suggested to female reformers that women's suffrage was one of many changes that could be expected in the near term as part of a larger movement of well-educated, rational individuals moving society onward and upward towards perfection.[53]

Spencerian social evolution, better known today as social Darwinism, enjoyed tremendous support among American intellectuals and reformers at the turn of the twentieth century.[54] While the high school history textbook narrative tends to link social Darwinism to robber barons and the interests of capital over labor that characterized Gilded Age society, historian Robert Bannister and others have shown that this was not exactly the case. As Bannister argues, "more intriguing than social Darwinism itself is what one might term the myth of social Darwinism—the charge, usually unsubstantiated or quite out of proportion to the evidence, that Darwinism was widely and wantonly used by forces of reaction."[55] At the turn of the twentieth century, Spencerian arguments were enlisted by reformers and reactionaries alike, often to contrasting ends. The promise of intentional, continuous progress appealed to Americans across the political spectrum who saw the turn of the century as a time of possibility and change, as evidenced by the wide array of reform movements that flourished during this time period.

Despite the fact that they had much in common, a fault line emerged between freethinking feminists and religious suffragists at least in part over whether or not embracing evolutionary theory also meant questioning biblical authority, a question that often hinged on which school of evolutionary theory one subscribed to.[56] This schism also represents one manifestation of the evangelical efforts described earlier in this chapter to align Darwinism with atheism and encourage people to choose between God and Darwin. Given this false choice, the Christian suffragists who made up the vast majority of NAWSA members chose God and social Darwinism, whereas the freethinking feminists, led by Stanton, continued referencing Darwin and critiquing biblical authority. Unlike Darwinian theory, social evolutionary theory did not directly confront the biblical creation story, nor did it press the concept of animal-human kinship or challenge the existence of a divine creator. Overall, social Darwinism did more to support than challenge the Anglo-American elite's way of life, making it possible for religious believers and nonbelievers alike to rally around it. As a result, social Darwinist theories did not encounter nearly as much backlash or religious opposition as did the materialistic Darwinian evolution, and they fit well within the overall rhetoric of progress promulgated by NAWSA and WCTU reformers.

The social evolutionary argument for women's rights was perhaps best articulated by Carrie Chapman Catt, the influential president of NAWSA from 1900 to 1904 and again from 1915 to 1920, the years of the final successful push for suffrage. Although Catt was not an enthusiastic or orthodox Christian herself, she helped formulate NAWSA's social evolutionary rhetoric. At the 1893 World's Fair in Chicago, Catt delivered a speech that signaled her enthusiasm for social evolutionary ideals. In "Evolution and Woman's Suffrage," Catt described evolution as "not an hypothesis but an absolute proof that the 'world does move;' that it moves ever onward and upward, that the path of men leads ever nearer and nearer to the perfect and ideal."[57] According to Catt, the only thing holding society back from further progress was that woman had not yet taken her rightful place alongside man at the ballot box, though she trusted this was on the near horizon because "evolution, the greatest truth discovered in our century, is on our side." By evolution, Catt did not refer to change over time by random variation and natural selection, nor did she mention Darwin, animals, survival, or mating in her address. She referred instead to the idea that the world inevitably progressed "nearer the perfect and ideal" and that with "work" women could have a greater say in government that had heretofore been corrupted by less virtuous men.[58]

Catt's version of evolution also buttressed her belief that white women were the most deserving of the vote, often at the expense of people of color and immigrants, because they were the most educated or the most "evolved." Historian Kevin Amidon notes that in Catt's arguments between 1902 and the winning of the vote in 1920, "race was continually linked to sex as an integral part of an evolutionary system of differentiation and evaluation," generally to differentiate between the needs of white, educated women and nonwhite others.[59] To Catt and many of her fellow reformers, middle- and upper-class white women deserved the same privileges as white men because of their shared levels of education and refinement, so they lobbied for "educated suffrage," which in practical terms meant white suffrage. In keeping with popular evolutionary thought of the day, they believed that Anglo-Saxons were simply more evolved than other races and that white women had been mistakenly grouped with other downtrodden people when in fact they deserved to be considered on par with white men. As the historian Louise Michele Newman and others have persuasively argued, white racial superiority was a core element of women's rights rhetoric, and women often invoked evolutionary discourse regarding the racial hierarchy of civilization to support suffrage arguments based on whiteness.[60]

Catt's beliefs about women and evolution also fit squarely with historian Beryl Satter's concept of "evolutionary republicanism." In the early republic, concerned citizens joined with national leaders in stressing that virtuous citizens—economically independent, educated, and moral—needed to sacrifice for the common good in order for the nation to survive. As economic independence, a key component of being a "virtuous citizen," became increasingly unavailable during the Gilded Age, Satter argues that middle- and upper-class reformers "found a new grounding for the virtue of the nation's citizens in beliefs about Anglo-Saxon evolutionary superiority."[61] While the Darwinian feminists were increasingly at odds with the NAWSA suffragists after 1890, the two groups both tended to draw on evolutionary discourse to advocate for votes for educated, native-born white women.

Here again, however, to label the Darwinian feminists, most notably Stanton, as intractably racist, as Louise Michele Newman and others have essentially done, overlooks the nuanced historical context in which these women lived and the range of arguments that they advocated. Certainly, as Kathi Kern has established, the freethinking Darwinian feminists were largely blind to their own racial and class privilege, which was "what allowed them to see gender as the source of all oppression."[62] Yet, at the same time, historians Ann Gordon and Michele Mitchell encourage a broader lens and a more nuanced appreciation of context and intellectual history when it comes to assessing the role of race in Stanton's thought and, by extension, the views of other freethinking Darwinian feminists. Gordon argues that historians who claim Stanton's views were racist and unchanging between 1869 and the mid-1890s "must ignore Stanton's core convictions and oversimplify complex problems in her thinking and in American history."[63] Gordon situates Stanton's support for educated suffrage within the larger political and reform climates in which she worked, establishing that Stanton never stopped believing in universal suffrage and that calls for educated suffrage were not necessarily made by reactionary racists. For example, by the 1890s many leading African Americans, including W.E.B. Dubois and the congressman John Mercer Langston, also supported education and literacy requirements for voting, especially "if applied to both races equally."[64] While disavowing the racist tone of many of Stanton's writings in the late 1860s, Michele Mitchell also urges careful attention to historical context and intellectual history. As she explains, "the debate over whether Elizabeth Cady Stanton was racist or merely elitist is not as relevant as is the sort of racial knowledge available to her during the volatile, early years of Radical Reconstruction. The context in which Stanton pushed for women's suffrage was one in which citizenship was partially reconfigured through race, in which race, gender, and class were germane to struggles over citizenship." In the late 1860s, and indeed for much for the second half of the nineteenth century, the popular evolutionary description of a racialized hierarchy from savage to civilized provided convenient language for Stanton to voice her disapproval of black (and other) male suffrage at the expense of white women's. Thus, Stanton's articulation of this racialized hierarchy can be understood as, in Mitchell's words, "at once intriguing, surprising, regrettable, contradictory, and predictable."[65]

Those scholars who have analyzed the ways in which women's rights activists utilized evolutionary rhetoric have tended to focus on its applications in terms of race and to argue that its influence was detrimental because it encouraged women to think in terms of difference, permanently deterring the movement from its egalitarian goals.[66] That Stanton's and other suffragists' rhetoric was often race based and in many cases racist, to modern readers if not to contemporaries, is most certainly true, echoing the predominant interpretation of evolutionary theory in American culture at the time. When it came to thinking about the differences between women and men, however, the turn to difference had implications in addition to those regarding race. Just as women voiced race- and class-based arguments in campaigns for educated suffrage, they also began articulating demands for women's rights grounded in the conviction that women were fundamentally different from men. A key component in their gendered thinking was that whiteness was essential to their gender, but they also began making innovative feminist arguments inspired by the ways in which evolution had demonstrated, at least in their minds, that women differed from men.

After decades of invoking the language of equality and natural rights to argue for women's inclusion in the body politic, many leading activists realized this strategy simply was not working and maybe even stopped believing in it themselves, especially after the crushing defeat of universal suffrage in Kansas and the failure to include women in the Fourteenth and Fifteenth Amendments, which granted emancipated male slaves, but not women, the right to vote. Antebellum female activists were inspired by claims of natural rights, and they frequently extrapolated Enlightenment principles to include women (as vividly established in the 1848 Declaration of Sentiments produced by attendees at the first women's rights convention, which was modeled nearly word-for-word on the Declaration of Independence), but these arguments repeatedly failed to convince male leaders, as well as the vast majority of Americans. Furthermore, as Ann Gordon has established, popular acceptance of suffrage as a natural right waned in the late 1860s as political leaders began legislating and writing about the vote as a "privilege," not a right.[67] Women's rights leaders, none more so than Elizabeth Cady Stanton, were keen to this shifting political landscape.

In September 1868, Stanton, author of the 1848 Declaration of Sentiments and a leading proponent of the inclusion of women in the Fourteenth and Fifteenth Amendments, devoted nearly an entire issue of her newspaper, the *Revolution*, to a speech on the "identity of the sexes in mind" delivered before the British Association for the Advancement of Science by Lydia Becker, a British suffragist, botanist, and correspondent of Darwin. Stanton applauded Becker's decision to discuss the "woman's sphere" from the perspective of "pure science," but she rejected Becker's argument that men and women were equal. As Stanton explained:

> We started on Miss Becker's ground [equality] twenty years ago, because we thought, from that standpoint, we could draw the strongest arguments for women's enfranchisement. And there we stood firmly entrenched, until we saw that stronger arguments could be drawn from a difference in sex, in mind as well as body. But while admitting a difference, we claim that difference gives man no superiority, no rights over woman that she has not over him. We see a perfect analogy everywhere in mind and matter; and finding sex in the whole animal and vegetable kingdoms, it is fair to infer that it is in the world of thought also.[68]

Writing in 1868, on the heels of the defeat of universal suffrage, Stanton realized the futility, at least in that historical context, of arguing for women's rights on the basis of equality with men. Furthermore, as the Darwinian feminists contemplated vast structural changes to society, they realized that men, as well as women, would have to change. Arguments for changing men's roles were harder to make using the language of equality. If men and women were equal, why then would husbands, male-dominated workplaces, and man-made laws have to change? Couldn't they simply be amended to allow women to do just as men did? While denying biological determinism and continuing to stress the structural and cultural elements of gender oppression, the Darwinian feminists' acknowledgement of some biological sex differences—maternity and breast-feeding, for example— allowed them to advance creative innovations and demands, as the following chapters attest.

Moreover, the concept of natural rights, while certainly revolutionary, was not gender neutral. As Thomas Laqueur and others have established, natural rights rhetoric not only left out women, it was expressly constructed to exclude them and eviscerate whatever small political and other privileges (wealthy) women may have had. In order for natural rights language to be persuasive in the seventeenth and eighteenth centuries, whole bodies of knowledge—science, medicine, philosophy—had to be rewritten to define women as fundamentally different from men and, thus, not deserving of natural rights.[69] As a result, by the nineteenth century, medical and popular opinion converged in viewing women as the polar opposites, rather than the mirror images, of men, making it hard for women to convincingly advance arguments on the basis of natural rights. After the failure to include women in the Fourteenth and Fifteenth Amendments and the defeat of universal suffrage in Kansas, many women's rights activists grudgingly came to understand that "all men are created equal" really did mean all "men." Yes, it was inspiring to think that "all men are created equal" might one day include women, but embedded in the very same Enlightenment ideology was the conviction that women could never be considered in the same category as men. Stanton and other women's rights leaders eventually accepted this paradox of Enlightenment philosophy and looked to new thought structures on which to ground their demands. Thus, I am suggesting that we reevaluate women's shift from equality to difference by focusing on the complex interplay among the Bible, natural rights, and evolutionary theory and ask why, given these options, some leading feminist thinkers chose to ally themselves with evolutionary science and difference.

In addition to strategic concerns about the effectiveness of equality arguments, it became harder for women to believe in natural equality after the scientific and popular embrace of Darwinian theory. In a Darwinian world, all organisms were not created equally. And these slight differences between organisms often decided who lived long enough to pass on traits to the next generation. Moreover, the differences between males and females, especially as described in *The Descent of Man*, provided the key to evolution. As Darwin explained, sexual dimorphism was found throughout most forms of organic life, and it enabled the evolution of higher animals (because it multiplied the possible variations that offspring could inherit). Thus, sex differences were both natural and vital. Darwin noted that "advancement or progress in the organic scale" rested "on the amount of differentiation and specialization of the several parts of a being," notably the differences between males and females that were a telltale feature of vertebrates.[70] Taken together, these insights encouraged women's rights advocates, along with much of the reading public, to think in terms of difference. To Stanton and other Darwinian feminists, admitting (some) biological sex differences did not mean acknowledging female inferiority; rather, it allowed them to critique the male world of work, religious orthodoxy, and politics, instead of simply asking for entry into it, and to suggest that men and male-dominated systems needed to change in order for their goals to be met.

## DARWINIAN FEMINISTS CHALLENGE
## THE LEGACY OF EVE

Women's rights activists from across the ideological spectrum tended to shift from equality to difference arguments, but the 1890 merger of NAWSA brought the tensions between the freethinking Darwinian feminists and the more conventionally religious social evolutionists to a head. A vocal minority of women, especially those who had been active in the antebellum feminist movement, rejected NAWSA's adoption of Christian rhetoric and opposed the merger of AWSA with NWSA.[71] These women felt strongly that a top priority for the movement was for women to sever themselves, once and for all, from associations with Eve and to affiliate instead with science. Matilda Joslyn Gage, for example, so opposed the 1890 merger that she formed her own short-lived organization, the Woman's National Liberty Union, which insisted upon the separation of church and state and rejected any church doctrine that taught that woman was a secondary creation. Gage, Elizabeth Cady Stanton, Helen Hamilton Gardener, and a core group of like-minded women believed that Christian doctrine provided the intellectual foundation for the oppression of women, and they steadfastly opposed any affiliation with orthodox Christianity, as well as any reforms based on the idea that the United States was inherently or exclusively a Christian nation. After decades of agitating for women's rights, these women had come to believe that Christian cosmology based on Adam and Eve was the single most powerful barrier to female equality.[72] In the words of Stanton, "it is on this allegory [the "petty surgical operation" that supposedly created Eve from Adam's rib] that all the enemies of women rest their battering rams, to prove her inferiority."[73] Evolutionary theory provided intellectual support for critiques of the rib story and scientific evidence for the simultaneous creation of men and women.

In 1893, Gage published *Woman, Church and State* in which she laid out her argument that "the most grievous wrong ever inflicted upon woman has been in the Christian teaching that she was not created equal with man."[74] Specifically, she charged the Christian church with forming and upholding the "patriarchate," which systematically oppressed women in all stages of life, while at the same time obscuring the history of the pre-Christian "matriarchate" in which women ruled. The state, drawing insight and guidance from the church, adopted similarly misogynistic policies against women. Again, the root cause of both church and state discrimination against women came down to the biblical creation story. Gage observed that Eve continued to govern women's place in modern society: "In nothing has the ignorance and weakness of the church been more fully shown than in its controversies in regard to the creation. From the time of the 'Fathers' to the present hour, despite its assertions and dogmas, the church has ever been engaged in discussions upon the Garden of Eden, the serpent, woman, man, and God as connected in one inseparable relation."[75] Gage was hopeful, however, that science would ultimately "free [woman] from the bondage of the church" by revealing "that Christianity is false and its foundation a myth, which every discovery of science shows to be as baseless as its former belief that the earth is flat."[76]

Like Gage, Elizabeth Cady Stanton became increasingly convinced that Christian cosmology, grounded in the Eve myth, mandated women's oppression. As early as 1863, Stanton made waves in the women's rights movement by declaring that "a book that curses woman in her maternity, degrades her in marriage, makes her the author of sin, and a mere afterthought in creation and baptizes all this as the word of God cannot be said to be a great blessing to the sex."[77] Throughout the 1860s and 1870s, Stanton's growing frustration with organized religion manifested itself in numerous articles in the *Revolution* and, later, the *Woman's Tribune*, where she honed in on the Bible as the ultimate source of women's degra-

Figure 1.3. Elizabeth Cady Stanton (c. 1866–1871), cabinet
photography by Napoleon Sarony. Photograph courtesy of
Special Collections, Fine Arts Library, Harvard University.

dation (fig. 1.3).[78] Stanton chided her shortsighted and, in her view, naive
colleagues for clinging to the Bible. Beginning in 1878, Stanton and her
core group of followers brought forth resolutions condemning organized
religion for the subordination of women at every NWSA convention. So
strong was their resistance to associations with Eve that the 1885 resolution proposed that NWSA disavow association with any religious body that
taught women were inferior as a result of creation.[79] By the 1880s, Stanton
had come to believe that women should sever ties with the church once
and for all. In an article published in the freethought and pro–birth control newspaper *Lucifer the Light-Bearer*, Stanton argued that the church
was built on the oppression of women and, thus, unlikely to acknowledge

"liberty for a sex supposed for wise purposes to have been subordinated to man by divine decree." Recognizing women's equality would "compel an entire change in church canons, discipline, and authority, and many doctrines of the Christian faith." Stanton concluded, "as a matter of self-preservation, the Church has no interest in the emancipation of woman, as its very existence depends on her blind faith."[80] As a result of such writings, Stanton, always an eclectic thinker and iconoclast, found her ideas becoming anathema to her peers who could see suffrage on the horizon and who did not want to diminish their chances of attaining the vote by being associated with someone who voiced such unorthodox views.

Ultimately, Stanton's anticlericalism led to her ouster from the women's rights movement that she had done so much to establish, and she found a new home among the freethinkers. At the end of the nineteenth century, the popularity of Darwinian evolution propelled freethought, a secular movement that spanned the continuum from agnosticism to atheism, from the fringes of respectability to the mainstream. According to Susan Jacoby's history of freethought, the period from 1875 to 1914 was the Golden Age of the movement, largely because evolutionary theory gave credence to its main claims.[81] Throughout the 1870s and 1880s, the most popular speaker on the national lecture circuit was Robert Ingersoll, the "great agnostic." Freethought was especially influential on those individuals, including Stanton and Gage, "who moved from liberal Protestantism to outright agnosticism."[82] Never one to espouse "any deep spiritual conviction," according to Stanton biographer Lori Ginzberg, Stanton "was far more impressed" with Darwin's theory of evolution than with any religious fads or ideas.[83]

Evolution provided a boon in adherents and respectability to the freethought movement which, in turn, helped spread innovative ideas on gender and provided crucial forums in which Darwinian feminists could publish and speak as the women's rights movement contracted to focus on the vote. Stanton became close friends with freethought leader Robert Ingersoll and his wife Eva, and she found a soul mate in fellow freethinking feminist, Helen Hamilton Gardener. Stanton also befriended Benjamin Franklin Underwood and his wife Sara, who popularized Darwin in the United States and published the freethought periodicals the *Index* and *Open Court*.[84] Unlike NAWSA, freethought groups welcomed critiques of marriage, traditional gender roles, and the Bible, and they, too, cheered Darwinian evolutionary theory for introducing the possibility of a completely materialistic universe. The freethought movement provided an ideal outlet for Stanton who was frustrated by continually having to debate clergymen and her more orthodox peers. As Stanton expert Kathi Kern contends, "it would be difficult to overestimate the importance of the freethought movement to Stanton personally, politically, and intellectually."[85] Throughout her life, Stanton sought a community that would support her far-ranging beliefs about women's rights, and, it seems, the closest she came was her fellow agnostics and atheists in the freethought movement.

Buttressed by her supportive community of freethinkers and by evolution's refutation of the very existence of the Garden of Eden, Stanton decided to make it her mission to convince women that the Bible and the biblical creation story in particular were responsible for their second-class status. To Stanton, clearing up the confusion resulting from Genesis was the linchpin in the broader campaign for women's rights. To aid in this endeavor, she latched on to evolutionary science because it provided the ideal ballast to fight the legacy of Eve. As Stanton wrote in *Lucifer the Light-Bearer* in 1891, "What would be the tragedy in the garden of Eden to a generation of scientific women?" "Scientific women" would "relegate the allegory to the same class of literature as Aesop's fables."[86]

Central to Stanton's critique of patriarchal religion was the time she spent in Europe in the early 1880s recovering from the strain of publishing volume 2 of *The History of Woman Suffrage* and visiting her children, Harriot Stanton Blatch in England and Theodore Stanton in France. During this time, she enjoyed the cosmopolitan secularism of Europe, read evolutionary theory, and became further convinced that suffrage could not elevate women unless they also freed themselves from the belief that Eve caused the fall of humanity and accepted that organized religion was predicated on their oppression.[87] While in London, Stanton confided in her diary that she had "dipped into Darwin's *Descent of Man* and Spencer's *First Principles*, which have cleared up many of my ideas on theology and left me more than ever reconciled to rest with many debatable ideas relegated to the unknown."[88] Or, as she wrote her cousin, Elizabeth Smith Miller, "Admit Darwin's theory of evolution and the whole orthodox system topples to the ground; if there was no Fall, there was no need of a Savior, and the atonement, regeneration and salvation have no significance whatever."[89]

Emboldened by her study of evolution and her time among European freethinkers, Stanton returned to the States determined to reveal the male bias at the heart of organized Christianity through the publication of the *Woman's Bible*, which she considered to be her greatest contribution to women's emancipation. Stanton explained the impetus for the *Woman's Bible* in the freethought newspaper the *Index*:

Believing that the source and centre of woman's degradation is in the religious idea of her uncleanness and depravity, as set forth with innumerable reiterations in the Old Testament . . . the [Woman's Bible] committee feel it to be their conscientious duty to investigate the authenticity of the Scriptures. If convinced that they emanate from the customs and opinions of a barbarous age, and have no significance in the civilization of the nineteenth century, they hope to free women from the bondage of the old theologies, by showing that The Bible rests simply on the authority of man, and that its teachings are unfit for this stage of evolution in which the sexes occupy an equal place in the world of thought.[90]

While Stanton was hardly a devout student of Darwin's (she also drew on Auguste Comte, Herbert Spencer, and Jean-Baptiste Lamarck), the popularization of Darwinian evolutionary theory freed her to interpret the Bible as allegory because it was definitive proof of the limits of biblical authority and, especially, because it provided scientific evidence for her contention that Adam and Eve never existed.

The *Woman's Bible* consisted of reprints of all the biblical passages relating to women, which according to Stanton made up just ten percent of the book, alongside commentaries written by Stanton and the revising committee members.[91] In these commentaries, the women focused on translation issues, biblical history, and textual analysis. The commentaries on Genesis provided the dramatic core of the text and were shaped by the writers' familiarity with evolutionary discourse. As Stanton explained, "Scientists tell us that 'the missing link' between the ape and man, has recently been discovered, so that we can now trace back an unbroken line of ancestors to the dawn of creation." Because the allegorical tale in Genesis enabled "the doctrines of original sin, the fall of man, and woman the author of all our woes, and the curses on the serpent, the woman, and the man; the Darwinian theory of the gradual growth of the race from a lower to a higher type of animal life, is more hopeful and encouraging."[92] To Stanton, having apes as ancestors, rather than Eve, boded well for women's rights.

The mainstream press initially greeted the *Woman's Bible* with curiosity, but most reviews of the work criticized either the quality of the writing or Stanton's temerity in selecting such a heretical topic. The only venues where her work was favorably and enthusiastically received were the freethought journals and the *Woman's Tribune*, which was published by *Woman's Bible* revising committee member Clara Bewick Colby.[93] Whether or not they agreed with Stanton's conclusions, however, numer-

ous reviews recognized that Eve continued to define women's place in society. The *Omaha World Herald*, for example, described the *Woman's Bible* as "Eve's version of that little Eden episode."[94] Some women who wrote
letters to the *Woman's Tribune* disagreed with Stanton's arguments, but
all conceded that the source of women's subjection could be found in the
Garden of Eden. One correspondent called it "the one great rock of ignorant superstition which, more than any other, blocks the road of woman's
progress."[95] The favorable review in *Lucifer the Light-Bearer* concurred
that the "rib story" was "unscientific," "unreasonable," and warranted
women's full attention. This reviewer saw grounds for hope in "sexual science," which "concerns the happiness and well being of our race" and "is
of more importance to man than a knowledge of any or of all other sciences
put together."[96] In addition to refuting Eve, evolutionary theory made this
sexual science possible, as subsequent chapters will demonstrate.

Not to be outdone by clerical and mainstream opposition to Stanton's
project, NAWSA and the WCTU also publicly criticized the *Woman's Bible*.
After having spent years promoting women's inherent piety and moral imperative to assume a larger role in church and public affairs, these women
did not want to be associated with Stanton's quixotic quest. The *Woman's Bible* caused so much controversy that Carrie Chapman Catt spear-
headed a movement to denounce the text at the 1896 NAWSA convention
in Washington, D.C. Only a handful of attendees had read the *Woman's
Bible*, but most felt that it damaged the cause by association and scared
away potential adherents.[97] Essentially, the debate over whether or not to
censure the *Woman's Bible* was also a debate about the future of NAWSA
and the larger movement for women's rights: were women solely interested
in the vote or did they also want larger, systemic changes? Furthermore,
what roles should religion and science play in the movement? After heated
debate, Stanton's defenders brokered a toned-down resolution, which read,
"That this Association is non-sectarian, being composed of persons of all
shades of religious opinion, and that it has no official connection with
the so-called 'Woman's Bible,' or any theological publication."[98] Charlotte
Perkins Gilman, perhaps the best-known Darwinian feminist, had the
mumps and was not planning to attend the 1896 NAWSA convention, but
she dragged herself from her sickbed so that she could "fight the resolution
disavowing the Woman's Bible." She proposed a counter-resolution that declared NAWSA's nonsectarian nature but did not specifically mention the
*Woman's Bible*; it failed by five votes. Ultimately, the resolution publicly
disavowing the *Woman's Bible* passed by a vote of fifty-three to forty-one.[99]
Within a few short years, Stanton and her freethinking colleagues, includ-

ing *Woman's Bible* revising committee member Lillie Devereaux Blake who ran for NAWSA president in 1900 against Carrie Chapman Catt, were purged from the movement and, for much of the next one hundred years, its history.[100]

While the *Woman's Bible* controversy established that the majority of suffragists were not prepared to follow Darwinian evolutionary theory to its most material conclusions, at least not publicly, it also proved that the days of thinking about gender solely in terms of the Bible were over. In this regard, Eve had given way to evolution. Regardless of where one fell on the orthodox–freethought continuum, evolutionary theory provided a new way for women to view the universe and their role in it, and a new language to describe what they saw. Evolution reframed the terms of gender debates from biblical ancestors to animal kin, from individual to species, and from piety to reproduction. Based more on women's bodies than on women's souls and more on women's biological function as mothers than on their religious faith, science, nevertheless, offered the promise of objectivity. In the 1870s and 1880s, a variety of women's rights activists eagerly enlisted evolutionary theory as an ally largely because it provided an alternative to the Genesis creation story and decentered Eve as the barometer by which all women would be judged. Ultimately, the most enthusiastic female adherents to Darwinian evolutionary theory were found outside of the organized suffrage movement, while those influenced by what the philosopher of science Michael Ruse has called "evolutionism" could be found everywhere.[101] After 1890, potentially blasphemous discussions of Darwinian evolution were banished from NAWSA, but the women writing before 1890 left an important and understudied legacy of Darwinian feminism; and, after 1890, feminist women and men working outside of NAWSA continued to probe the revolutionary implications of thinking about gender and sex in terms of evolutionary science. The following chapters chronicle the lives and writings of many of the women who voted against sanctioning Stanton's *Woman's Bible* and who joined Stanton in opposing the NAWSA merger—women whose views of emancipation were informed by evolutionary science and included much more than the vote.

As NAWSA coalesced to work on behalf of women's right to vote and to promote Christian women's influence on society, other reform groups—freethought, socialist, and sex reform—continued to agitate on behalf of women's complete emancipation from male dominance—in the professions, in the classroom, in the church, and in the home. In these circles, Darwinian evolutionary theory was widely discussed and often seen as supporting expanded opportunities for women, more equitable domestic relationships, and increased sexual autonomy for women. Even though women were confronted with an onslaught of scientific and medical studies proving their "natural" inferiority in the 1870s and 1880s, many welcomed this change of base, as chapter 2 will demonstrate. Unlike the halls of Congress or the inner sanctums of churches, women were players in the evolutionary saga, and they, too, could study the latest evolutionary science, interpret it against their experiences, and challenge the experiments of others. Of course, biblical calls for and against women's rights persisted, but, after the publication of Darwin's *The Descent of Man* in 1871, the major forum for debates about women's rights began to shift from Genesis to *Popular Science Monthly*. Women's tactics reflected this change—a change that in many ways they had helped to engineer.

## CHAPTER TWO

# "The Science of Feminine Humanity"

In July 1925, when suffrage leader and feminist author Helen Hamilton Gardener (born Alice Chenoweth) died, her last will and testament contained an unusual stipulation. Gardener specified that upon her death, her brain be removed from her body and donated to the Burt Wilder Brain Collection at Cornell University for scientific examination, as long as it was intact and not damaged by disease. To ensure that her wishes were carried out to the letter, Gardener twice amended her will, named the specific doctor who would remove her brain, and consulted various lawyers to make sure the document was airtight. Like many of her peers, Gardener believed that one's level of intelligence and aptitude could be determined by looking at one's brain. Unlike most of her contemporaries, however, Gardener rejected the idea that brains differed according to sex—or at least that her brain had been hampered by her sex. She believed that her brain represented the highest development possible for a woman and, thus, that it would provide the ideal specimen to compare with the often-studied brains of eminent men. As she explained in her will, she hoped her donation would "aid science in making a fairer comparison between the brains of men and women 'who think.'"[1] She also believed, as this dramatic bequest demonstrates, that science should be the ultimate arbiter of questions regarding sex difference.

In her seventy-two years of life, Gardener wrote eight books, delivered countless addresses on the reform lecture circuit, married two times, traveled the world, and helped secure the passage of the Nineteenth Amendment by orchestrating behind-the-scenes talks with high-ranking congressional and White House officials, including President Woodrow Wilson and Speaker of the House James Beauchamp "Champ" Clark. Although Gardener had opposed the 1890 merger of the National Woman Suffrage

Fig. 2.1. Helen Hamilton Gardener, c. 1920.
Photograph courtesy of the Library of Congress.

Association with the American Woman Suffrage Association to form the
National American Woman Suffrage Association (NAWSA), she became
a NAWSA vice-president after moving to Washington, D.C., in 1907.
Throughout the 1910s, when NAWSA leaders needed to deliver a sensitive message or plead a point to a powerful elected official, they enlisted
Gardener, whom they gratefully referred to as the organization's "diplomatic corps."[2] Upon the ratification of the Nineteenth Amendment in
1920, President Wilson wanted to send a signal that women now held a
prominent place in the nation's affairs. So he nominated Gardener to fill
a vacancy on the U.S. Civil Service Commission, making her the highest
ranking woman in federal government (fig. 2.1). Women's rights activists
cheered her selection, and Gardener relished her high-profile post. When
she died just five years later, however, she believed that her biggest accomplishment was yet to come: Gardener wanted the scientific dissection

of her brain to establish, once and for all, that women's brains were not structurally, or otherwise, inferior to men's. Such claims were a staple of nineteenth-century antifeminism, but Gardener trusted in experimental science, interpreted objectively, and she believed that once a brain like hers—educated, well trained, and active—had been studied, no one would dare make such ill-informed claims again. She was partially right.

Helen Gardener's brain donation demonstrates the extent to which many nineteenth-century women trusted in science as an ally, as well as the extent to which debates about women's rights often hinged on women's bodies. This extraordinary story also represents one chapter in a longer struggle, led by women, to redefine mind-body dualism. Darwinian evolutionary theory provided compelling evidence for the idea—popularized by Rousseau and others during the Enlightenment—that physical traits correlated with mental ones. Since women's bodies tended to be smaller than men's, then surely their intellectual capacities were smaller too, or so the thinking went. In the nineteenth century, women conceded that there were physical differences between the sexes, but they struggled against the notion that these physical differences indicated any sort of mental (or other) inferiority. Furthermore, this little-known episode epitomizes women's efforts to help shape the emerging discipline called "science" and utilize it as a feminist tool, efforts that peaked in the 1870s and 1880s and were subsequently abandoned as the women's rights movement contracted to focus on the vote and as science became increasingly professionalized and masculinized.[3] In fact, by the time Gardener made her historic brain donation, it received front-page coverage for several days in the *New York Times* but scant attention in the women's press. Gardener began merging her commitment to feminism with her interest in science in the 1870s and 1880s, by engaging in highly publicized debates over the structure of women's brains, and perhaps did not realize that she was a relic in this regard by the 1920s, or perhaps she wanted her final statement on women's rights to remind younger generations of the symbiotic relationship between science and feminism.

By the mid-1870s, if one wanted to influence the debate on women's rights, it was no longer enough to consult the Bible—one also needed to be armed with scientific, ideally evolutionary, evidence. As Gardener trenchantly observed, "manly men are beginning to blush when they hear repeated the well-worn fable of the fall of man through woman's crime and her inferiority of position and opportunity, justified by priest and pleaser, because of legends inherited from barbarians—mental deformities worthy of their parentage." But, Gardener warned, "Conservatism, Ignorance, and

Egotism" had called in "medical science, still in its infancy, to aid in staying the march of progress." As a result, "Equality of opportunity began to be denied to woman, for the first time, upon natural and so-called scientific grounds . . . It was no longer her soul, but her body, that needed saving from herself."[4] Key women's rights activists recognized the shift from religious to scientific denials of women's equality, and, during the 1870s and 1880s, they resisted inaccurate pronouncements about their abilities veiled in the discourse of "science" and began trying to shape the emerging field of science themselves. In the words of Antoinette Brown Blackwell, an outspoken Darwinian feminist, in order to refute the false pronouncements of male scientists, women had to create the "science of feminine humanity."[5]

Rather than being passive victims of the new science of sex differences, women actively participated in this science by demanding to have their experiences objectively studied and by critiquing the biased methods of male scientists.[6] Unfortunately, these female science enthusiasts, much like the white male scientists they critiqued, generally turned a blind eye to analogous examples of racism in the emerging natural sciences, leaving an unjust and unfortunate legacy of scientific racism for the twentieth century. This chapter chronicles the developments leading up to Gardener's historic brain bequest, and it tells the larger story of women's engagement with science during the final third of the nineteenth century, a trend many attributed to the popularity of Charles Darwin and the questions his work raised about the practice of science and the biology of sex differences.

## "TO TEACH THE TRUTH IN NATURE"

Gardener's brain donation was the most dramatic example of women using their bodies and their physical experiences to create a more accurate and inclusive biology of sex difference, but she was not the only woman to recognize the radical potential of science, especially evolution, for those interested in women's rights.[7] Indeed, over one hundred women were so inspired by evolutionary science that they corresponded with Darwin himself.[8] Hundreds more published science-related articles in magazines including *Popular Science Monthly*, enrolled in summer science classes such as those offered by Harvard, or participated in discussions of scientific topics sponsored by women's clubs.[9] Many more read about scientific discoveries and the accomplishments of female scientists, which were front-page news in the *Woman's Journal* and the *Woman's Tribune* throughout the 1870s and 1880s.[10] One of the earliest and most prolific advocates of women's engagement with science was Antoinette Brown Blackwell.

Antoinette Brown Blackwell knew firsthand what it felt like to be barred from intellectual and professional goals because of her sex, and, like her acquaintance Helen Gardener, she ultimately looked to science for recourse. When Blackwell entered Oberlin College in 1845, she intended to become the nation's first ordained female minister, despite the fact that her advisor, the legendary revivalist preacher Charles Grandison Finney (and pretty much everyone else), disapproved of women speaking in public, to say nothing of a woman leading a congregation.[11] On September 15, 1853, she succeeded, ascending to the pulpit of her own Congregationalist parish in South Butler, New York. However, her hard-won and historic tenure lasted only a few months. After all the years of fighting the church and educational establishment for the right to preach, Blackwell began to lose her faith in Christian orthodoxy.[12] While she retained her belief in an omnipotent higher power, she officially resigned from her pulpit on July 20, 1854, and turned her attention to science.[13] In 1869, Blackwell published her first book, *Studies in General Science* (and sent a copy of it to Darwin), and in 1875 she became the first woman to publish a feminist critique of evolutionary theory with *The Sexes throughout Nature*.[14] In her later years, Blackwell struggled to combine her belief in a higher power with science by writing about the scientific basis of life after death; though she was no longer an orthodox Christian, Blackwell's definition of science involved a heavy dose of spirituality.[15] Blackwell's earlier writings highlight the ways in which feminists responded to evolutionary theory, and they document the shift that occurred in the second half of the nineteenth century as women and men increasingly looked to evolutionary science, rather than the Bible, to better understand sex differences and women's proper role in society. Or, as Blackwell put it, "it is time to recognize the fact that the 'irrepressible woman question' has already taken a new scientific departure."[16]

Blackwell did not have any formal scientific training, except for some youthful scientific investigations conducted with her brother, nor did she administer any scientific experiments herself.[17] Rather, she read widely in evolutionary science and attempted to reconcile what she read with what she saw around her and what she experienced as an educated woman and mother. She also exhibited a characteristic nineteenth-century enthusiasm for science and the scientific method. Blackwell biographer Elizabeth Cazden describes Blackwell's firm faith that "the scientific method of rea-

soning from established facts" would ultimately lead to a true understanding of gender difference, as well as determine whether or not there was an
afterlife.[18]

While Blackwell was enthusiastic about science, she had her doubts
about the authority invested in the findings of male scientists. As Blackwell insisted, "[I]t is to the most rigid scientific methods of investigation
that we must undoubtedly look for a final and authoritative decision as to
woman's legitimate nature and functions." Whatever the results, she implored women to "most confidently appeal" to "Nature as umpire—to Nature interpreted by scientific methods."[19] The problem, according to Blackwell, was that scientific methods were often perverted by male scientists.
In particular, Blackwell lamented that evolutionary theory had been misinterpreted by "the wisest, the highest, the most progressive and the most
influential authorities in science to-day." Because they were "standing on
a learned masculine eminence, looking from their isolated male stand-
points through their men's spectacles and through the misty atmosphere
of entailed hereditary glamour," these authorities could see only evidence
of women's "natural" inferiority.[20] Especially egregious in this regard were
Charles Darwin and Herbert Spencer, "thinkers who have more profoundly
influenced the opinions of the civilized world than perhaps any other two
living men. . . . and [who] endorsed by other world-wide authorities, are
joined in assigning the mete and boundary of womanly capacities." Unfortunately, Darwin and Spencer accepted the theory of "male superiority" as
a "foregone conclusion" rather than establish it scientifically using "adequate tests" and "careful and exact calculation."[21] Furthermore, Blackwell
fumed, what exactly did men like Darwin and Spencer know about "the
normal powers and functions of Woman"?[22] The problem, then, was not
science but the fact that science was being conducted mainly by men, for
men, and did not include accurate studies of women.

According to Blackwell, if one really wanted to learn about women,
one must turn to women themselves. Expecting a backlash, Blackwell admitted, "I do not underrate the charge of presumption which must attach
to any woman who will attempt to controvert the great masters of science
and of scientific inference." "But," she claimed, "there is no alternative!
Only a woman can approach the subject from a feminine standpoint; and
there are none but beginners among us in this class of investigations."[23]
What women lacked in specialized training and laboratory access, they
made up for by having female bodies and female experiences, traits which
no male scientist could boast. Woman, Blackwell advised, "must consent
to put in evidence the results of her own experience, and to develop the

scientific basis of her differing conclusions." If woman failed to speak out, she must "forever hold her peace, consent meekly to crown herself with these edicts of her inferiority."[24] Understanding the differences between men and women, according to Blackwell, required "a deeper reading of facts, a reconsideration of all the old data, from the bottom upwards; in a word, a new science—the science of Feminine Humanity." As Blackwell explained, the key to this new science was that "the experience of women must count for more here than the observation of the wisest men."[25] This science was not only "new" because it studied women firsthand; it was also "new" because it challenged the masculine boundaries being erected around the practice of science itself.

Throughout the 1870s and 1880s many of the nation's leading women joined Blackwell in her efforts to create a "science of feminine humanity." In the summer of 1886, for example, Smith College, the prestigious women's school in Northampton, Massachusetts, erected an historic building: the Lilly Hall of Science, the nation's first building dedicated to women's scientific studies and experimentation. Founded in 1872 as a bequest from Sophia Smith to provide women the "means and facilities for education equal to those which are afforded now in our Colleges to young men," Smith College quickly became a leader in the higher education of women and the first to offer women the standard male curriculum.[26] Central to this challenging curriculum was science. In his inaugural address, Smith president L. Clarke Seelye explained that the college wished to avoid "that narrowness which has always been the bane of female education" and, instead, to encourage young women to study the natural sciences so that they would be prepared to "feel an interest in the progress of science; to clearly comprehend its important discoveries, and to be prepared to make, afterward, in some chosen field, original investigations."[27] This was a bold undertaking. At the time, few believed that women were capable of comprehending science, let alone conducting original investigations.

Smith students were especially interested in evolutionary science and the fields of zoology and biology, but by the early 1880s the young women's interest in science had outpaced the college's infrastructure. President Seelye endeavored to find a donor to fund the construction of a building dedicated to scientific study among women, which proved a difficult task. In 1884, he happened to share a ride to Boston with Alfred T. Lilly, a wealthy, iconoclastic entrepreneur from nearby Florence, Massachusetts, who eventually offered his financial support.[28] Lilly had made his fortune in silk manufacturing, and he was a supporter of women's education, as well as a critic of Christian orthodoxy.[29] Seelye recalled that Lilly had told him that if the funds had been needed for a male institution, he "would never give a cent," but he "believes in science, and believes that truth is as valuable for women as men." According to Lilly's wishes, the engraved plaque on the Lilly Hall of Science reads, "Gift of Alfred Theodore Lilly to teach the truth in nature."

By all accounts, the female students of Smith relished the new laboratory spaces made available to them. As Smith student Gertrude Gane wrote her mother in 1893:

> I enjoy thoroughly my lessons this term, particularly Zoology. It is simply fascinating. We have nine laboratory hours a week as well as two lectures. We are now working on the skeleton of vertebrates, and I have already manipulated a mud puppy and an alligator (small). Today I spent about five hours in the laboratory, preparing a beautiful great rat. In the midst of the operation, I cooked him, and the savory odors were greatly enjoyed by all the students.[30]

So popular were the laboratory classes that by the 1890s, just a few years after the dedication of Lilly Hall, the students were already clamoring for additional laboratory space. Steel and railroad tycoon Andrew Carnegie agreed to provide half the funds for a new biology lab building in 1905, but the building was not completed until 1914.[31] For twenty-eight years, Lilly Hall was the seat of science at Smith College and, indeed, a model for the nation, as figure 2.2 shows.

The ramifications of Lilly's donation were both symbolic—building laboratories for women showed that they could contribute to scientific progress, not just learn about the innovations of others—and practical, as generations of female students availed themselves of its state-of-the-art facilities. Reporting on this landmark donation, the *Woman's Tribune* called it "magnificent" and reprinted long excerpts from Lilly's speech about the importance of scientific education for women.[32] The *Woman's Journal*, the official paper of the American Woman Suffrage Association, also devoted front-page coverage to this historic occasion, noting that applications to Smith were on the rise and that the next entering class would likely be the largest yet.[33] The Lilly Hall of Science concretized women's burgeoning interest in science and the growing consensus among those interested in female advancement that science was good for women.

At the same time, the women's club network—which included book clubs and volunteer societies, as well as more overtly feminist groups—also sought to engage with science. Most notably, the Association for the

Fig. 2.2. Smith College students in Professor Harris Hawthorne Wilder's zoology class laboratory, Lilly Hall of Science, Smith College, c. 1895. Photograph courtesy of the Smith College Archives, Smith College, Northampton, Massachusetts.

Advancement of Women (AAW), which was founded as a national organization for professional women, did much to promote science among women.[34] Although smaller in number than the larger reform organizations such as the Woman's Christian Temperance Union, AAW members tended to be very prominent members of their communities. Member Anna Garlin Spencer described AAW membership as "a union primarily of achieving personalities."[35] An outgrowth of Sorosis (the first organization for professional women) and the New England Women's Club, the AAW convened national congresses from 1873 to 1897. Their events attracted women from a variety of professions, although the women themselves tended to be white and middle or upper class, and their agendas tell us much about what was on "achieving" women's minds at the end of the nineteenth century. One item at the top of the AAW members' priority lists was science.

Maria Mitchell, the first AAW president and a noted astronomer, was perhaps the most dedicated and influential advocate for women in the sciences. In 1847, Mitchell discovered a comet that bears her name, and she was elected to the American Academy of Arts and Sciences in 1848. When famed naturalist Louis Agassiz sponsored her for membership in the American Association for the Advancement of Science in 1850, the members unanimously approved her application. Throughout her life, Mitchell

passionately lobbied on behalf of women in the professions, especially science. In her presidential address at the 1875 AAW convention, Mitchell
expounded on the need for women in science: "In my younger days, when
I was pained by the half educated, loose, and inaccurate ways which we
all had, I used to say, 'How much women need exact science,' but since I
have known of some workers in science who were not always true to the
teachings of nature, who have loved self more than science, I have said,
'How much science needs women.'"[36] Like Antoinette Brown Blackwell
(her colleague on the AAW's science committee), Mitchell seemed to suggest that scientific investigation could be conducted more accurately, free
from inherited privilege and bias, by women. Compared with men, she believed that women also had keener "perceptions of minute details" and
the "capacity for patient routine," which would be of "immense value in
the collection of scientific facts." Further, Mitchell believed that women's
daily activities prepared them for careers in science: "when I see a woman
put an exquisitely fine needle at exactly the same distance from the last
stitch . . . I think what a capacity she has for astronomical observations."[37]
To promote women's entry into science, Mitchell suggested that the AAW
found a science society where women "engaged in the study of natural or
physical science" could present their findings.[38] With this proposed AAW
science society, Mitchell hoped to set up an alternate path to professionalization for women, but the historical record does not indicate that this
society ever materialized.[39] She did note favorably, however, in the 1875
Science Committee report, the "very encouraging fact" that "women are
learning to give money to aid schools of science—for women as for men."[40]

　　Perhaps because of Mitchell's influence as the group's founding president, the AAW prominently featured scientific addresses at its annual
congresses. Indeed, scientific topics accounted for between one-third and
one-fourth of all papers delivered at AAW conventions held between 1873
and 1890.[41] At the 1875 AAW congress, for example, science committee
member Grace Anna Lewis provided a how-to guide for women interested
in science. She listed the schools that offered science courses for women,
the organizations women could join to learn about science (including the
American Philosophical Society and the American Academy of Natural
Sciences), and various museums around the country that women could
visit to learn about science.[42] To the women of the AAW, "science" implied being modern, thinking independently, and being able to understand
the natural world. This sort of critical thinking was precisely what opponents of women's equality feared, and they called in their own version of
science to thwart women's advancement.

# THE SCIENCE OF GENDER AND
# THE GENDER OF SCIENCE

While many of the nation's leading women were convinced that science was good for their cause, male scientists, together with most mainstream Americans, were not so sure. Bolstered by the popularity of evolutionary theory, male scientists and physicians seemed obsessed with studying the female physique to pinpoint the "natural" basis for women's physical and intellectual inferiority. The resulting debates between male scientists and proscience feminists raised as many questions about the practice of science as they did about the biology of sex difference. As the historian Daniel Patrick Thurs persuasively argues in *Science Talk: Changing Notions of Science in American Popular Culture* (2007), "science" as a field and as a term was very much in flux at the end of the nineteenth century. Prior to 1850, science was more or less synonymous with any type of "systemized knowledge" and did not conjure larger meanings of a unique method, experimentation, or professional training.[43] The public reception of Charles Darwin provided a tremendous occasion to discuss what exactly science was and who should be doing it. Between 1870 and 1900, experts and laypeople debated whether science should present "just the facts" or whether science should also pose original questions, analyze data, test speculative hypotheses, and aggregate information into meaningful patterns.[44] By 1910, expert and popular notions of science solidified as science came to be understood as a specialized form of knowledge, practiced by trained professionals, following established protocols and methods, in university- or government-sponsored laboratories or studies. As Thurs emphasizes, however, these developments were contested, and laypeople, together with practicing scientists, helped create the meaning of science. Women, too, played a key role in this process, even though the emerging field of science largely excluded them.

As science gained in cultural authority, scientific practitioners, together with some of the reading public, engaged in what Thomas Gieryn has called "boundary-work" to distinguish science from not-science.[45] Boundary-work refers to the ideological and rhetorical practices employed by scientists to confer prestige, cultural authority, and intellectual autonomy upon science by separating it from other forms of knowledge.[46] Whereas before 1870, amateur studies were welcomed as valuable contributions to science—in fact, Darwin himself relied on countless amateurs to collect and share the data that made up his major works—after 1870 a distinction began to be made between amateurs and professionals. Thurs's

research further demonstrates that at the turn of the century "a variety
of institutional structures emerged that set researchers in scientific fields
apart as a professional class, protected their autonomy, determined correct procedure, and moderated disputes by sanctioning some kinds of
knowledge as real science."[47] By the 1910s, this boundary-work was nearly
complete as a clear line separated scientists from amateurs and scientific
research from other types of scholarship. Such boundary building had profoundly gendered and racialized implications.

At the same time that Americans were debating what exactly counted
as science, a closely related question was, Who could participate in science? Before 1850 there were no specialized university or graduate-level
programs in science, and the word "scientist" did not come into popular
parlance until the twentieth century.[48] As specialized science education
programs developed, however, they largely excluded women and African
Americans, who were also not welcomed into professional organizations,
nor were their "amateur" studies and experiments considered "science."
The general public, including practicing amateur scientists like Blackwell,
could access, read, and contribute to popular scientific periodicals, such
as *Popular Science Monthly*, but had less and less access to the professional scientific journals such as *Nature*. In 1891, Blackwell published
an article in a popular periodical entitled "Women in Science," in which
she observed that, compared with the gains women were making in other
fields, progress in science presented unique challenges. Women were very
interested in science, a "pre-eminently modern" field, but "the doors of
instruction generally not being open to them, it has been extremely difficult to climb up in some other way without being regarded as thieves and
robbers." Blackwell recounted the many successes of notable women scientists, including the astronomer Maria Mitchell and the botanist Mary
Treat, but she lamented that owing to a lack of access to "expensive appli-
ances of well-furnished laboratories, facilities for making difficult experiments and tests, skilled assistants, [and] the emulation and approbation of
co-workers" women had not yet been able to discover "great facts, laws or
principles which mark an epoch in science."[49] The Lilly Hall of Science
bucked these trends, but it stood out as an outlier, not the norm. Most
women interested in science in the late nineteenth century had to carve
out alternative spaces for themselves to learn about, comment on, and in
a few cases practice science through the women's club network, popular
periodicals, libraries, and museums. Such spaces offered women opportunities to critique the work of male scientists and help shape public under-

standing of science, but they did not tend to confer scientific authority or prestige on women themselves.

To the women writing enthusiastically about science in the 1870s and 1880s, however, it was not necessarily apparent that they were soon to be excluded from official science. To them, science meant freedom from stories about virgin mothers and evil temptresses; and science represented a burgeoning field of study that, when conducted properly, could reveal essential truths about nature and about people. Evolutionary science in particular appealed to these women because it implied that there was a connection among women, men, and all other living things as well as an orderly, knowable process explaining human development. When women engaged in science, they contemplated questions about the natural universe and their own bodies in systematic ways using experiential knowledge, experimental evidence, and survey data from other women, and they shared information about their bodies and their health with others, even if, for the most part, they were not allowed access to the newly forming scientific establishment.

Whereas the women examined in this chapter tended to invoke science to resist the status quo and present alternative possibilities, men within the scientific establishment generally invoked science to defend the status quo, at least in terms of gender. When it came to divining the natural order of things, most male scientists presumed that the status quo was natural—they only needed to explain how and why it had come to be. The women outside the burgeoning scientific establishment and the men within it often consciously realized that they had different definitions of science, and, in fact, the essays and experiments studied in this chapter underscore these cleavages just as they helped to establish what science meant at the end of the nineteenth century. The gulf separating women's hopes for science from the reality of science as it was then practiced also explains what drove Helen Gardener to donate her brain to Cornell nearly forty years after the brain size debates made headlines.

## IS INTELLIGENCE A SECONDARY SEX CHARACTERISTIC?

Nowhere was the link between the emerging definition of science and the future of women's rights more apparent than in the debates regarding the higher education of women, which raged throughout the 1870s and 1880s. "If [a woman] applies for admission to Harvard," Antoinette Brown Blackwell observed, "Harvard can offer its most humane denial in the name of Physiology."[50] Which is, more or less, what Harvard did. But, when faced with scientific studies claiming to confirm their physical and mental inferiority, women responded with their own scientific studies and with the evidence of their experiences. Previous examinations of the debates regarding women's education and intellect covered in this chapter have tended to focus on the misogynistic bias at the root of male pronouncements about female inferiority. These studies tell the story of how nineteenth-century scientists and doctors colluded to pathologize menstruation, essentialize women according to their maternal function, and, in general, bar women from the professions.[51] If one focuses on the writings of Edward Clarke, William Hammond, and, to a lesser extent, Darwin, this is certainly the story that emerges. However, if one also looks at the ways in which women disputed these theories and if one examines the long-term trajectories of these debates, a more nuanced story comes to light. In each case, women asserted their own definition of science that depended upon inclusivity and freedom from ideology, amassed their own data, and rejected biological determinism by challenging the supposed boundary between nature and culture. In most cases, women debunked pseudoscientific theories of female inferiority and demanded more rigorous and accurate descriptions of female physiology, expanding the scope of scientific knowledge and honing its practice.

In the decades following the Civil War, young women's realities differed tremendously from their predecessors, even those who had come of age only a decade or two earlier. Because the war took the lives of so many men, there were fewer men to enroll in college, fewer men to work the new jobs created by the vastly expanded postwar industry, and fewer men to marry. As a result, more women than ever before had to find ways to support themselves. At the same time, colleges and universities, smarting from rapid expansion and dwindling numbers of male students, began encouraging women—new sources of tuition—to enroll. Lacking the traditional option of early marriage and excited by the increasing opportunities available to them, a growing number of women pursued college degrees. In 1870, when just one percent of the population attended college, twenty-one percent of attendees were women. By 1910, women made up forty percent of college attendees.[52] These dramatic demographic changes elicited heated public debates. Was it healthy for women to attend college? Would doing so imperil future offspring? Opponents of female higher education argued that education not only dismantled sex differentiation but also stymied the evolutionary process by diverting women from motherhood. Furthermore, they argued, women simply were not suited to the rigors of higher education. For evidence, one needed only to look at their bodies.

The concurrent fascination with evolutionary discourse offered one compelling way to interpret sex differences. As one scientist noted, since Darwin "remodeled" natural history, it has "been found capable of throwing valuable lights, previously little anticipated, upon topics quite unconnected with the origin and attributes of zoological or botanical species." In particular, this author suggested that concerned citizens enlist evolutionary theory to mediate debates about women's proper role in society.[53] *The Descent of Man* (1871), in particular, framed the debate over the higher education of women in important ways. First, the theory that physical structures correlated with mental ones, and the concomitant idea that women's mental inferiority could be read on their bodies, owed much to the ways in which scientists and laypeople interpreted Darwin's work. In *The Descent of Man*, Darwin was primarily concerned with the origin of secondary sex characteristics, though he made several influential statements about the intellectual differences between the sexes. For example, he explained that over the course of many generations, male-versus-male competition for female mates, together with the male's greater participation in the struggle to survive, had forced men to develop more complex and varied skills than women, who simply waited to be charmed and protected. "There can be little doubt that the greater size and strength of man," explained Darwin, were due to eons of males competing "in the general struggle for life and in their contests for wives."[54]

To evolutionists, not only were women's hips designed for the production of offspring, so too were their thoughts and emotions. George Romanes, Darwin's friend and advocate, explained that "the maternal instincts are to woman perhaps the strongest of all influences in the determination of character."[55] So different were the resulting male and female intellects that Romanes suggested "in the animal kingdom as a whole the males admit of being classified, as it were, in one psychological species and the females in another."[56] To Romanes, women's inferior intellect was not a flaw but rather an evolutionary necessity for the creation of healthy offspring. Writing in *Popular Science Monthly*, M.A. Hardaker, a female author, concurred that since maternity took up "twenty percent of the energy of women between twenty and forty years of age," intellectual equality was not a goal to be pursued because it would lead to the extinction of the species.[57] With stakes this high, it is no wonder that female education was a defining women's rights issue in the 1870s and 1880s.

Women's intellectual capacities, according to Darwin and most other evolutionists, were permanently limited by their reproductive functions, which drew the lion's share of their energy and of evolutionary attention.[58] Thus, over many thousands of years, Darwin concluded, "man has ultimately become superior to woman." "It is, indeed, fortunate that the law of the equal transmission of characters to both sexes has commonly prevailed throughout the whole class of mammals," Darwin noted, "otherwise it is probable that man would have become as superior in mental endowment to woman, as the peacock is in ornamental plumage to the peahen."[59] Darwin further believed that, technically speaking, women could possibly be educated to an equal intellectual standing with men over many generations but at too great a cost to the "easy education of our children" and the "happiness of our homes."[60] Even though Darwin himself supported female education, opponents seized upon the peahen quote to argue that educating women went against nature's plan and was ultimately futile, if not injurious.[61]

Darwin's thinking about female education was also indebted to French evolutionist Jean-Baptiste Lamarck (1744–1829). Lamarck's major contribution to evolutionary thought was his theory that traits acquired in one's lifetime, including temperance and intellectual capacity, could be passed on to one's offspring, a theory often referred to as "Lamarckianism." His ideas, although frequently contested, remained plausible until the turn of the twentieth century when they were discredited by the experiments of August Weismann. Many experts, and even more laypeople, believed that habits and acquired traits could be transmitted to the next generation, thereby making education an obvious vehicle for those who wanted to tailor the evolutionary process to fit their goals. The Lamarckian model of heredity also helps explain why evolutionary scientists were so interested in the question of female education: it would be one thing to educate a few exceptional women but quite another to simultaneously improve women's lot for eternity.

Furthermore, evolutionists contended that the extent to which males differed from females, in both physical traits and day-to-day activities, corresponded to their level of evolutionary advancement. In *The Descent of Man*, Darwin asserted that sex differences promoted the evolutionary process by efficiently dividing labor and that the most advanced species were those in which the sexes were the most differentiated. As evolutionists saw it, animals progressed from asexual to sexual reproduction, developing increasingly complicated mating systems as they ascended the evolutionary ladder.[62] At the very top of this ladder were those humans with the most strictly defined gender roles: married couples in which the husband worked outside the home and the wife tended to children and domestic tasks, couples that also tended to be middle or upper class and white. To those men steeped in evolutionary discourse and the attendant pride in being at the pinnacle of all living things, women going to college threatened to minimize sex differentiation, thwart evolutionary advancement, and diminish white racial superiority.

## "A FAIR CHANCE FOR THE GIRLS"

Scientific objections to the higher education of women reached a crescendo with the publication of Dr. Edward H. Clarke's *Sex in Education, or a Fair Chance for the Girls* (1873).[63] Clarke was a well-respected ear and eye doctor, a Harvard professor, and a member of Harvard's Board of Overseers. He had previously made comments in defense of a group of beleaguered female medical students in Pennsylvania and, as a result, the New England Women's Club invited him to deliver an address. The clubwomen thought they had invited an ally to speak on the subject of "women's fitness for entering practical life." Likewise, Clarke anticipated a friendly audience before whom he could unveil his theory that higher education unfit women for motherhood and made them sick. Both sides thought wrong. Although Clarke had defended the female medical students against the boorish behavior of their male colleagues, he did not think women's bodies could withstand the pressures of college. Clarke's presentation "on the health of women, as affecting steady, persistent mental application" was followed by a heated debate during which a majority of the women challenged his views about the connection between higher education and female illness.[64]

To clarify and expand his points, Clarke published *Sex in Education*, which became one of the most frequently debated and influential works of the 1870s, drawing attention from scientific and medical authorities, the popular press, and women's rights activists.[65] By linking female malaise to female accomplishment, and by tying both to evolutionary progress, this book helped set the tone for debates about the science of sex difference for decades. Clarke's book was nothing short of a national phenomenon. *Sex in Education* went through seventeen editions in thirteen years; it was reviewed in prestigious national periodicals, including the *New York Times*, the *Nation*, and *Popular Science Monthly*; and countless women read it or were evaluated according to its standards. At the newly coeducational University of Michigan, two hundred copies reportedly sold in one day. Future Bryn Mawr President M. Carey Thomas recalled the anxiety of going to college in the age of Edward Clarke: "We did not know when we began whether women's health could stand the strain of education. We were haunted in those days by the clanging chains of that gloomy specter, Dr. Edward Clarke's *Sex in Education*."[66]

Clarke was inspired to turn his attention away from eyes and ears and towards female physiology by the debate over whether or not to admit women to Harvard, which raged during the early 1870s. Like many of his colleagues, Clarke opposed women's entry into Harvard's classrooms. He based his objections to female education on the Darwinian worldview that sex differentiation was essential to evolutionary progress. As he explained, "differentiation is nature's method of ascent."[67] If men and women were to participate in similar activities and lead similar lives, then the species—or, more specifically for Clarke, middle- and upper-class white people—would suffer. Clarke's antidote to "identical co-educational" institutions, including state universities in the West and women's colleges in the East, was to design an educational system that accentuated sex differences in both structure and content. Specifically, Clarke recommended overhauling educational practices to suit the "periodicity" of females and the "persistence" of males by separating the sexes, limiting the number of hours per day that girls could study, and forcing girls to take off every fourth week of school to coincide with their menstrual cycles.

Clarke grounded his arguments on the popular idea that female menstruation necessarily brought with it decreased mental capacity and energy. According to Clarke's model, women had to choose between developing themselves as individuals and the ability to give birth to healthy offspring. As Clarke explained, "the muscles and the brain cannot *functionate* in their best way at the same moment."[68] Any strenuous mental exertion during girls' developmental years came at the expense of their reproductive potential. If women persisted in seeking higher education during their formative years, Clarke warned that a third gender would evolve: a sexless woman, which he named "agene" and equated with "the sexless class of termites."[69] As evidence, Clarke drew on his observations of women and related anecdotes from male colleagues.

Feminist men and women recognized the threat *Sex in Education* posed to female advancement and organized a powerful counterattack that reshaped debates about women's physiology and about how science should be practiced. For starters, Clarke's opponents pointed out that the ramifications of his plan extended far beyond schools because he defined education broadly as "comprehending the whole manner of life, physical and psychical, during the educational period."[70] What was at stake in these debates, then, was not just female admission to college but whether or not women could pursue any interests outside the home. As Drs. George and Anna Manning Comfort pointed out in their response, Clarke's plan would ultimately dismantle female education because "it would be impossible to organize schools in which every pupil is to refrain from study, or from class exercise, for from four to seven successive days in each month."[71] Without an educated female labor pool, job opportunities would be greatly diminished and women would be further tethered to home and hearth. The danger of *Sex in Education*, according to Eliza Bisbee Duffey, one of Clarke's most trenchant critics, was that this book "is more than it seems to be. It is a covert blow against the desires and ambitions of woman in every direction except a strictly domestic one."[72]

*Sex in Education* also prompted women's rights activists to question what exactly constituted scientific evidence and who could speak authoritatively about women's bodies. Many of the nation's most famous women rallied in opposition to Clarke. At least four books, one novel, and dozens of articles and speeches refuted *Sex in Education*.[73] Many questioned his methods and demanded more evidence; others thought that he had misunderstood menstruation or had no business talking about it in the first place. In response to what they considered to be Clarke's outrageous and dangerous assertions, women demanded more female physicians, the right to speak for themselves and their bodies, and a verifiable scientific account of sex difference, not the cobbled together compilation of secondhand anecdotes that Clarke offered.

The most common and effective response to *Sex in Education* was to question Clarke's evidence and call for more studies. Ironically, it was Darwin who provided Clarke's detractors with the gold standard of exacting scientific research. Prominent women's rights activist Thomas Wentworth Higginson led the charge against Clarke in the pages of the *Woman's Journal*, observing, "Darwin offers his basis of facts as modestly and as amply as if he were an unknown man; and proceeds step by step, still fortifying himself, or stating frankly where he is unfortified." This was in contrast to Clarke, who "by no means comes up to the recognized standard of science either in the quantity or the quality of the facts on which he bases his argument."[74] The "standard of science" was still being worked out in the 1870s, but to many readers it was clear that Clarke's methods were not satisfactory.

Antoinette Brown Blackwell helped orchestrate the response to *Sex in Education* and craft the feminist approach to evolutionary science more broadly. To Blackwell, women's involvement in science did not necessarily mean conducting laboratory experiments or attaining scientific degrees, although that was certainly one way women could be involved. What she really wanted was for women to keep abreast of the latest scientific research and measure it against their own standards. Did scientists' statements about women accord with women's own experiences? If not, then women needed to correct the record and speak for themselves. She trusted that women's case would be safe with science, as long as science included the voices and experiences of women. To counter Clarke, she cited her own twenty-four years as a student in coeducational facilities and the fact that she had always been in good health.[75] The *Westminster Review* lauded Blackwell's section on Clarke in *The Sexes throughout Nature* as "the most convincing," largely because her experiences as a college graduate and mother spoke for themselves.[76]

Galvanized by *Sex in Education*, many women heeded Blackwell's call to speak authoritatively about their bodies, countering Clarke's litany of female malaise with their own stories of good health. Instead of relying on doctors to speak for them, women queried female college graduates or wrote about their own lives. Nearly all of the respondents noted never feeling healthier than they did in college.[77] Elizabeth Cumings, for one, argued that education kept women mentally and physically healthy and helped them avoid hysteria and other mental disorders.[78] Similarly, many university officials claimed that Clarke misrepresented the situation at their schools where women were doing just fine, in body and in mind. In addition, the *Woman's Journal* published testimonies from college professors, administrators, health professionals, and female graduates affirming that, on the whole, college women were healthier than their less-educated peers and that, if anything, education and exercise kept them that way. In contrast to the indirect and often anonymous testimony that Clarke provided, these first-person accounts proved a potent weapon.

Other women rejected the crass, reductive way in which their bodies and lives were dissected by someone with no firsthand experience.[79] In addition to her direct response to Clarke, *No Sex in Education*, Eliza Bisbee Duffey also wrote an advice book for women that emphasized her female perspective.[80] According to Duffey, "men have had their say. It is but fitting that a woman should have hers, especially as the woman who assumes to speak does so with an authority man cannot venture to claim." As a woman, wife, and mother, Duffey felt that she had superior qualifications to those "possessed by any man, professional or otherwise." Further emphasizing this point, Duffey used a variant of the word "woman" three times in the title of her advice book: *What Women Should Know,*

*A Woman's Book about Women.* Duffey wrote the book because she had become "exasperated" when "these champions [of woman's sphere] insist upon making this weakness of mind and body constitutional—something inherent in the sex."[81] "Can a natural state be called a state of invalidism?" asked Duffey. Healthy women, she argued, experienced no distress during menstruation and "should themselves decide as to their capabilities."[82] She hoped her advice book would be a first step in this process.

What finally undercut Clarke's argument about the taxing nature of menstruation was the historic research conducted by Dr. Mary Putman Jacobi, the pioneering female physician and lecturer at the Woman's Medical College of the New York Infirmary for Women and Children. Owing to the popularity of Clarke's book, Harvard University chose the following question as the topic for its prestigious Boylston essay competition in 1876: "Do women require mental and bodily rest during Menstruation; and to what extent?"[83] In fact, the Boylston Prize committee questioned Clarke's research methods and hoped that the contributed essays would shed more reliable light on this important topic.[84] Entrants submitted their essays anonymously, and Jacobi recognized the potential for a judicious outcome. With the help of her colleagues, she surveyed hundreds of women about their menstrual cycles, levels of exercise and activity, and amount of suffering. She then submitted the essay, "The Question of Rest for Women during Menstruation," to the competition. Out of the 268 women who completed her survey, ninety-four reported being "completely free from discomfort during menstruation" and twenty-eight said that they suffered only slightly or occasionally. Overall, a majority of respondents did not experience significant discomfort during menstruation. Based on these extensive surveys—by far the largest of the time—Jacobi concluded that most women did not suffer during menstruation and that those who suffered the least were the most active, both physically and mentally. Conversely, the women most likely to suffer menstrual pain were those with little formal education or those enrolled in "ornamental" education, such as finishing school. Jacobi concluded, *"There is nothing in the nature of menstruation to imply the necessity, or even the desirability, of rest, for women whose nutrition is really normal"* (italics in original).[85] Jacobi won Harvard's Boylston Prize, soundly discrediting Clarke's thesis that women needed to rest while menstruating, as well as his anecdotal methods and moralistic tone.

Jacobi's study of menstruation was also a foray into the ongoing cultural conversation regarding the definition of science. According to Jacobi biographer Carla Bittel, Jacobi "consciously asserted her technical achievements and condemned the failures and inadequacies of her rivals. She directly confronted work she deemed unscientific, especially the work of the infamous Edward Clarke."[86] In a preliminary paper she wrote in response to Clarke, Jacobi discredited Clarke's findings by linking them to ideology, not science. As she noted, Clarke's theory "appeals to many interests besides those of scientific truth."[87] In her Boylston Prize–winning essay, Jacobi convinced the Harvard judges by using "statistics, diagnostic laboratory tools and the science of nutrition," which she correctly assumed her learned audience would value.[88] Thus, as Bittel persuasively argues, Jacobi's victory in the Boylston Prize contest did not signal the prize committee's support for women in higher education as much as it indicated that they shared her definition of science as laboratory and experiment based.

Following suit, the scientific and medical community, by and large, responded positively to Jacobi's essay, especially because it epitomized the emerging consensus about how science should be practiced. Subsequently, many other researchers copied her methods, but most rejected her conclusions about the healthfulness of women's higher education and about menstruation. According to Bittel, Jacobi felt slighted by the response to her prize-winning essay and was "very disappointed that a large body of her work, specifically her subsequent articles on menstruation, had been ignored." Nevertheless, her research shaped the parameters of scientific practice and laid the foundation for the science of feminine humanity that she and other leading women had long imagined. Though she did not live to see it, Jacobi remained convinced until her death in 1906 that "science would one day lead to social emancipation" for women.[89]

## THE BRAIN SIZE DEBATES

As a woman who took science classes at Columbia University in the 1870s and engaged in feminist activism in the 1880s, Helen Hamilton Gardener was surely familiar with the Clarke debates and with Jacobi's research. While living in New York City, Gardener also befriended freethought leader Robert Ingersoll, the "Great Agnostic" and the most popular speaker on the lecture circuit. Ingersoll encouraged Gardener's ambition and atheism, and she embarked on speaking tours of her own. Known as "Ingersoll in Soprano," Gardener railed against the sexual double standard and criticized the Christian church for fostering the subjugation of women.[90] Gardener was not actively involved in the women's rights movement, however, until one of her essays caught the attention of suffrage leaders. As a student of science, Gardener had become incensed by the popular theory that women's intellectual inferiority could be read on their supposedly smaller and less developed brains. In 1887, she turned her attention to convincing the public, especially women themselves, that women's brains were in no way inferior to men's and that female physiology did not limit women's mental powers.

When Mary Putnam Jacobi refuted Clarke's theory about the taxing nature of menstruation, the debate over female education shifted from the symptoms of women's periods to the size and structure of their brains. If menstruation did not explain women's intellectual inferiority, then surely their smaller, less developed brains did. The year after *Sex in Education*, Edward Clarke published a book elaborating on British physician Henry Maudsley's theory that sex differences were evident in brains as well as bodies. In *The Building of a Brain* (1874), Clarke slightly modified his recommendations from *Sex in Education*. Now he suggested that girls, as well as boys, not study more than six hours a day and that domestic and technical education be interfered with only in "exceptional cases." While menstruating, "all girls would require a periodical remission of variable length, from the labor of physical education, such as gymnastics, long walks, and the like; and also all would require a remission from the labor of social education, such as dancing, visiting, and similar offices."[91] Noticeably absent from this list of activities to be avoided during menstruation were study and mental exertion. His book, however, aimed to convince readers that brains were indeed sexed and needed to be developed along separate male-female educational tracks. According to Clarke, male brains were charged with the "command [of] a ship;" women's brains with the "govern[ance] of a household."[92] Fundamentally, then, Clarke's argument in *The Building of a Brain* mirrored that of *Sex in Education*, though he gave women slightly more leeway to study and refrained from disparaging comments about manly spinsters.

For evidence about the female brain, Clarke drew heavily on the work of Dr. William A. Hammond, a prominent neurologist with a keen interest in the differences between male and female brains. After treating scores of injured soldiers as Surgeon General of the U.S. Army during the Civil War, Hammond focused his professional attentions on diseases of the mind and nervous system.[93] He founded the American Neurological Association and later served as its president. Through his research on nervous disorders, he became convinced that there was a link between female education and mental breakdown. As he explained to Clarke, "It falls to my lot to see a good many young ladies whose nervous systems are exhausted, and thus rendered irritable, by intense application to studies for which their minds are not suited." He recalled in particular one young female patient who "was compelled to study civil engineering and spherical trigonometry,— subjects not as likely to be of use to her as a knowledge of the language of Timbuctoo." Schools such as hers, Hammond charged, "do more to un-sex women than all the anomalies who prate about the right to vote, and to wear trousers."[94] For evidence Hammond, like Clarke, drew heavily on anecdotes of patients he had seen or heard about from colleagues, not on blind laboratory studies or surveys judiciously interpreted.

Throughout the 1870s and 1880s, as Gardener studied science and de-livered fiery speeches on the lecture circuit, Hammond honed his theory about the differences between male and female brains. Ultimately, he de-termined that female brains were structurally different from, and inferior to, male brains in nineteen distinct ways, including weighing on average five ounces less (Hammond claimed "the larger the brain the greater the mental power of the individual"), displaying less distinct convolutions, and possessing thinner gray matter.[95] These differences in brain struc-ture, according to Hammond, explained women's failure to attain intel-lectual or professional prominence. He argued that "grave anatomical and physiological reasons demand not only that the progress of [the women's rights movement] should be arrested, but that, contrary to the ordinary course of procedure in other revolutions, this one should go backward."[96] Women had advanced beyond what their inferior brains were capable of handling. While women's brains were "perfectly adapted to the proper sta-tus of woman in the established plan of nature," such brains "would in-evitably make the worst legislator, the worst judge, the worst commander of a man-of-war."[97] In short, women were intuitive not abstract, imitative not original, and emotional not reasonable. Such descriptions of female intellect were common in the nineteenth century, Darwin himself said as much, but Hammond was the first to link female inferiority to the struc-ture of the female brain.[98] Ten years after women beat back Clarke's biased studies of menstruation and education, they were faced with an analogous argument based on the weight and structure of their brains.

For as long as Hammond had been expounding on the inferiority of the female brain, women had been responding to him in the pages of popu-lar journals and from the podiums at women's rights conferences.[99] Ham-mond had raised their ire not only through his statements about brain structure but also because of his opposition to women's suffrage and his characterization of women's rights activists as "short-haired women and long-haired men" who were "disappointed in their efforts to get husbands or wives, or else unhappy in their domestic relations."[100] Throughout the

1870s and 1880s, Antoinette Brown Blackwell, Elizabeth Cady Stanton, and others denounced Hammond at every opportunity.[101] Blackwell, for instance, argued that men's brains were bigger only because they needed to control men's larger bodies and that women made up for their smaller brains by having more complex nervous systems.[102] Stanton pointed out that scientific descriptions of women's brains lacked the scrupulous attention to experimentation and method that characterized other scientific work.[103] Others contended that if brain size did indicate intelligence, elephants would be the leaders of men and giants would rule the planet. At the same time, scientists promoted analogous arguments about the "inferior" brains of African Americans and other people of color, though the women who critiqued Hammond failed to make this connection.[104]

Debates about female brains did not take center stage, however, until 1887, when Hammond delivered a speech entitled "Brain-Forcing in Childhood," which was subsequently reprinted in *Popular Science Monthly* and numerous other periodicals.[105] Hammond's main point in "Brain-Forcing in Childhood" was that all students, regardless of gender, were forced to study too much and learn too many subjects at the same time. This system, he claimed, was especially pernicious for girls. Girls, according to Hammond, should stick to learning subjects to which they would be naturally called as mothers. For evidence, he cited the "comparative anatomy and physiology of the male and female brain." Key to his argument was the idea, grounded in Darwinian evolutionary theory, that the more advanced the species, the more distinct roles between male and female. Among humans, Hammond noted that "the skull of the male . . . is of greater capacity than that of the female, and it is a singular fact that the difference in favor of the male increases with civilization."[106] Thus, if women's brains were to evolve to be more like men's, this would actually be an evolutionary setback.

Helen Hamilton Gardener read this address and picked up her pen. She responded to Hammond in the pages of *Popular Science Monthly*, sparking several months of back-and-forth debate in the letters to the editor section. Their exchange highlights the contested status of nineteenth-century science and the high stakes for women in determining how science would ultimately be defined and practiced. Much like Mary Putnam Jacobi's response to Clarke, Gardener objected to Hammond's methods as much as to his sexist findings—findings that she suggested were based on "assumption and prejudice" rather than "scientific facts and discoveries."[107] To Gardener, Hammond's arguments were particularly dangerous because they carried the cultural authority of coming from a nationally respected scientist: "the writings of such a man, aided by the circulation and prestige of the leading journals of the country, which publicize them as authoritative, must inevitably influence school directors, voters, and legislators and go far to crystallize the belief that facts are well known to the medical profession, with which it would be dangerous to trifle."[108] But trifle she did.

Unable to conduct experiments on human brains herself, Gardener tested Hammond's findings by submitting a list of questions to twenty of the nation's top brain specialists, all of whom referred her to the leading expert on brain anatomy, Dr. Edward C. Spitzka, a prominent neurologist and anatomist in New York City. Having "previously discovered that even brain anatomists are subject to the spell of good clothes," Gardener put on her "best gown" and requested a meeting with the notoriously elusive and short-tempered Dr. Spitzka. Spitzka was impressed with the thoroughness of her questions and with the topic of her query, and the two struck up a vibrant exchange that formed the basis of her rebuttal to Hammond.[109] In particular, Gardener asked Spitzka if brain anatomists could identify the sex of individuals simply by looking at their brains (Hammond worked the opposite way—he knew the sex of the brains he studied and then asked what were their distinctive features). Since Hammond placed such emphasis on the size and structural differences between male and female brains, Gardener thought this would be a logical test of his theory. What prompted Gardener to investigate the claim that "there were natural anatomical differences between the brains of the sexes of the human race" was that no one made similar claims about the brains of "lower animals."[110] A firm believer in evolution, she found it incongruous that the brains of humans would develop so unlike those of other species. As a further test of nature versus nurture, Gardener asked whether brain specialists noted structural differences among infants' brains. Spitzka and the other experts informed her that they could not possibly determine the sex of an infant's or an adult's brain simply by looking at it. By establishing that scientists could not distinguish male and female brains by sight, Gardener hoped to bolster the idea that if any sex differences in brains existed, they were cultural, not biological.

In her letters in *Popular Science Monthly*, Gardener also stressed that, logically, Hammond's arguments made no sense. If, as he claimed, men's brains became increasingly advanced as civilization progressed, then men's brains were clearly benefiting from cultural changes—not biology. Thus, it made no sense to deny women access to these same cultural resources on the grounds that their brains "naturally" could not handle higher education. Furthermore, she noted if the differences between male and female brains were natural, they should be present in all races and groups of people, not only in the most civilized as Hammond claimed. To give him the chance to prove his point once and for all, Gardener proposed a challenge: if Hammond could successfully determine the sex of twenty brains she provided for him, borrowed from the collections of her brain anatomist friends, she would forever rest her case. Hammond replied that this challenge was preposterous and suggested, instead, that he provide her with twenty thumbs and ask her to identify the sex of the person from whose hand they came. The editors of the *Woman's Tribune* cheered Gardener from afar, declaring that if Hammond did not accept her challenge "we want to hear nothing more from him on the subject of woman's inferiority."[111]

The letters between Gardener and Hammond demonstrate the extent to which the emerging standards of scientific practice, above and beyond brain anatomy, formed the crux of the debate as each tried to establish that the other was not scientific enough. Throughout their sparring, Gardener positioned herself as the voice of reason and a force for truth but also as a self-conscious outsider to the scientific establishment. She did not mention her scientific courses at Columbia, nor did she reference published scientific studies that might have contradicted Hammond's or mimic Hammond's scientific tone. Rather, she wrote clearly, almost lawyerly, relying on logic and classical rhetoric to critique Hammond's claims, point by point. In his reply to Gardener's brain challenge, Hammond mocked her twenty leading brain specialists, insinuating that they were imaginary, and he critiqued the tone of her letter for its "unscientific spirit." Throughout his responses, Hammond cited his insider knowledge of brain anatomy and his familiarity with previous research and other researchers, and he chided Gardener for not being a member of the club, so to speak. Gender, of course, was an important subtext of his attacks. For example, he criticized Gardener's "feminine" proclivity for using italics and noted that she displayed the "defective logical power" so characteristic of female minds.[112] Gardener responded with more evidence and logical rebuttals to Hammond's anecdotes.[113] In closing, Hammond lamented having given Gardener more attention than she deserved and advised Gardener and her brain anatomists that "before they again rush into print they make themselves to some extent acquainted with the elementary truths of the science of anthropology."[114] In response, Gardener strenuously objected to the claim that anthropology was a science. A distrust of anthropology and anthropologists—the main source of evidence on which Hammond drew (including the work of Carl Vogt and Paolo Mantegazza)—pervaded Gardener's replies. As she pointedly noted in her second letter to Hammond, "[S]ince the science of anthropology is as yet in its infancy; since its various students disagree; and since within the past few months one of its cardinal principles has been found to be unsound, I am all the less willing to accept the sweeping statements of Dr. Hammond."[115] Gardener intimated that she had more faith in the hard sciences, those involving laboratory study and dissection, rather than in the soft sciences, which relied more heavily on the investigator's observations and secondhand anecdotes, although in the 1880s the distinction between anthropology and the experimental sciences was not yet firmly established, and Darwin, too, drew on many anthropological studies.[116]

To Gardener, science meant impartial interpretation of experimental results, which was why she thought her brain identification challenge provided the ideal rebuttal to Hammond's questionable claims. But this challenge raised a separate methodological question: should women and men be evaluated as members of a group or as individuals? Hammond refused the brain identification challenge on the grounds that it was impossible to distinguish differences between individual brains—what mattered were the aggregate differences between men and women. As a counteroffer, he suggested that they weigh one hundred male and one hundred female brains and then compare the averages, a test he was confident would prove that male brains weighed, on average, more than female brains. Relying on averages, not individuals, he noted confidently, was the way "all such determinations are made by those who know what they are about." He then lobbed a final challenge to Gardener: he listed several eminent men whose brains each weighed more than fifty-six ounces, daring her to find even one female brain that had ever surpassed this mark.[117]

Gardener did not engage Hammond's suggestion that they compare average brain weights, but she took on his individual brain weight challenge in earnest. She wrote a popular essay detailing her critique of Hammond entitled "Sex in Brain," which she was invited to deliver at the 1888 International Council of Women held in Washington, D.C., to commemorate the fortieth anniversary of the Seneca Falls Women's Rights Convention. The world's leading women's rights activists attended, and, despite the fact that Gardener had not previously been involved with the movement, the organizers allotted her a keynote spot, testifying to the extent to which the women prioritized science.[118] Gardener did not disappoint.

In her address, Gardener delivered a powerful argument about the possibilities and limitations of science. Women, she exhorted, "had hailed science as their friend and ally" only to be met with "pseudo-science" that "adopted theories, invented statistics, and published personal prejudices as demonstrated fact."[119] Gardener further suggested that science as practiced by many male scientists simply dressed up age-old religious ideas of female inferiority in modern scientific language, explaining that she had found that "a man's religious leanings inevitably color and modify all of his opinions, and govern his entire mental outlook."[120] Some of the brain specialists she contacted for her study performed "mental gymnastics" to make it seem as if their scientific findings adhered with their religious beliefs and "gave a black eye to their facts in preserving a blind eye to their faith."[121] Nothing, according to Gardener, could taint a man's scientific practice more than a belief in the Genesis creation story. Orthodox believers, no doubt, considered "'Adam as a creature after God's own heart and in his image,' and therefore capable and deserving of all opportunity and development for and because of himself, and to promote his own happiness." Whereas Eve became a "mere bone or rib of contention as it were, between man and man." "The more literal and consistent his faith," charged Gardener, "the less likely is he to deal with woman as an intellectual being, capable of and entitled to the same or as liberal, mental, social, and financial opportunities or rights as are universally conceded in this country to be the birthright of man."[122] The problem then was not science but science improperly practiced owing to the lingering influence of the Genesis creation story and its insistence on inherent female inferiority.

In contrast to men like Hammond who employed questionable methods and who let religious ideology taint their scientific research, Gardener held up Edward Spitzka as the epitome of a modern scientist. Spitzka, she explained, had in his laboratory "brains from those of a mouse to those of the largest whale on record." Gardener was also impressed with Spitzka's laboratory equipment that enabled him to show her "the peculiarities of brains as shown by microscopes and scales," and she appreciated that he "looked up points in foreign journals to which I had not access." Perhaps most important to Gardener, Spitzka demonstrated the ways in which empirical science could be helpful to women. Spitzka, she noted, "does not himself believe in the equality of the sexes, but he is too thoroughly scientific to allow his hereditary bias to color his statements of facts on this or any subject." Gardener concluded, "in the hands of a man who has arrived at that point of mental poise and dignity, our case is safe, no matter what his sentiments might be."[123]

Despite Gardener's hope in science as a vehicle for promoting women's rights, she stopped short of encouraging women to engage in scientific study or research themselves. She saw the boundaries of specialization and academic credentials barring women's access to scientific research, noting that few women "had the anatomical and anthropological information to risk a fight on a field which assumed to be held by those who based all of their arguments upon scientific facts, collected by microscope and scales and reduced to unanswerable statistics."[124] Instead, she implored women to be more informed and critical consumers of scientific knowledge, to question what they read, and to distinguish the good scientists from the bad. And, like Blackwell, she encouraged women to contribute their experiences and their bodies to science to make sure that the scientific record represented them.

What most troubled Gardener about Hammond's argument was that scientists had yet to study the brains of any remarkable women. Instead, they compared the brains of anonymous women who had died in state hospitals, or on the streets, with those of statesmen, writers, and other men of international renown. To even the scales, she implored her peers at the 1888 women's rights convention to consider donating their brains to science. "I sincerely hope that the brains of some of our able women may be preserved and examined by honest brain students, so that we may hereafter have our Cuviers and Websters and Cromwells," intoned Gardener. "And I think I know where some of them can be found without a search-warrant—when Miss Anthony, Mrs. Stanton, and some others I have the honor to know, are done with theirs."[125]

Elizabeth Cady Stanton heeded the call. After hearing Gardener's "Sex in Brain" speech, Stanton declared, "The paper read last night by Helen Gardener was an unanswerable argument to the twaddle of the scientists on woman's brain. The facts she gave us were so encouraging that I started life again this morning, with renewed confidence that my brain might hold out a few years longer."[126] This meeting solidified an intimate and sustaining friendship between Stanton and Gardener, two outspoken freethinkers. They supported each other's agnosticism and remained close friends and allies until Stanton's death.[127] They also took an important oath together: Stanton and Gardener pledged to each other that upon their deaths they would donate their brains to science so that, for the first time, researchers might compare the brains of eminent women with those of eminent men.

When Stanton died in 1902, she had indeed planned for her brain to go to Cornell University's Burt Wilder Brain Collection for dissection, and her signed brain bequest form remains in the Cornell archives. In 1887, Gardener had sent her a note, signed "Heathen Helen," asking her to donate her brain and convince her family to honor her bequest. On the back of this note, Stanton instructed her children: "you must save my brain for Heathen Helen's statistics."[128] Gardener publicly explained Stanton's wishes in a memorial address: "Mrs. Stanton asked me, in case she should go into the silence before me, if I would speak for her—at her grave. . . . First, she wished it known that she died as she had lived, a fearless, serene agnostic." Gardener lauded Stanton's decision to donate her "tireless brain" to Cornell University "that it might serve Science and mankind in helping to arrive at the truth, after death, as it always had done in life." According to Gardener, Stanton "felt that a brain like hers would be useful for all time in the record it would give the world, *for the first time*,—the scientific record of a thinker among women." Stanton hoped that her brain would contribute to "the fine heritage of all women" and be her "last and holiest gift."[129] But Stanton's heirs balked at this request and denied, mistakenly, that their mother had ever agreed to donate her brain to science.[130]

In the early 1900s, the idea that brain size corresponded with intelligence came under sustained attack. While skeptics had pointed out flaws in the theory in the nineteenth century, consistent empirical data to discredit the brain-weight theory of intelligence did not emerge until the social sciences, particularly psychology, took up the question. Historian Cynthia Eagle Russett credits the work of Alice Lee, then a graduate student studying under Karl Pearson at the University of London, with creating a formula for establishing skull capacity and then applying this formula to enough skulls, those of male anatomists who had volunteered, to establish that there was no clear link between brain size and intellectual capacity. Lee published her findings in 1902, and her mentor followed up with similar studies that same year.[131] Then, in 1909, Johns Hopkins anatomist Franklin Mall applied new statistical measures to the study of the frontal lobe and fissures of the brain, areas that had long been associated with both racial and sex differences. Mall found no differences between male brains and female brains, concluding, "[T]he general claim that the brain of woman is foetal or of simian type is largely an opinion without any scientific foundation." He further elaborated that any assertion "regarding male and female types are of no scientific value."[132] Mall's research and tone indicated that any supposed differences between male and female brains had come from the assumptions of male scientists rather than the female brains being dissected. But, still, no prominent woman's brain had been examined.

When Gardener died in 1925, she was a widow without children or other meddlesome heirs to derail her plans. Within hours of her death at

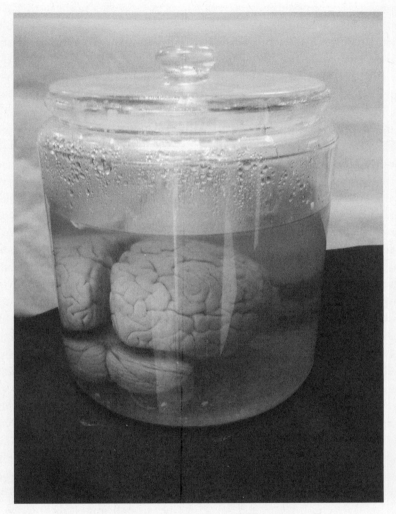

Fig. 2.3. Helen Hamilton Gardener's brain on display at Cornell University, 2013. Courtesy of the Burt Wilder Brain Collection, Department of Psychology, Cornell University. Photo credit: Sheila Ann Dean.

Walter Reed Hospital, Army Major Frank D. Francis packaged her brain and shipped it to Cornell's brain collection, where it remains on display today (fig. 2.3).[133] In her will, Gardener explained that in 1897 Burt Wilder, the founder of the Cornell brain collection that bears his name, had invited her to submit her brain as a "representative of the brains of women who have used their brains for the public welfare" and that after having spent her life "using such brains as I possess in trying to better the conditions of humanity and especially of women" she was happy to grant this request.[134]

Dr. James Papez, the curator of the brain collection in 1925, dissected Gardener's brain and published his exhaustive findings in the *American Journal of Physical Anthropology*. To modern readers, his fifty-page report contains so many measurements, comparisons, and qualifications—he literally dissected every millimeter of Gardener's brain—that his whole project seems absurd. In other words, it was exactly the kind of empirical study that Gardener criticized Hammond for not conducting and that she hoped might one day be the norm. Among the many figures recounted in his detailed description of Gardener's brain, Papez found that her brain had a lower "precuneal index" than the average female, whose precuneal index is lower than the average male's, but that she had a "remarkably high" occipital index. Ultimately, however, "the differences in the size of the medial frontal region in the two sexes is about .9 and is not sufficient to explain the difference that exists between the precuneal and occipital indexes of the two sexes."[135] But to Papez the message was clear.

In this maze of measurements, in which some women exceeded some men some of the time, but not always, and vice versa, Papez determined that Gardener's brain was in fact highly developed, in correspondence with her many achievements, and that "sex differences [were] exhibited to a lesser degree than in other female brains."[136] He did not proclaim the absence of sex in brain; he simply found Gardener's brain to be less sexed than those of the forty other female brains in his collection. He also noted that her brain weighed fifteen hundred grams, which "must be considered reasonably high for a woman whose normal weight was 106 pounds," even though it was just under the fifty-six ounce mark of true greatness that Hammond had set in 1887. Papez included several mentions of Gardener's brain weight, perhaps sensing that many readers would want to know, but he clarified that "no great value can be assigned to brain weight alone," and he made no mention of Hammond.[137] Throughout the article, Papez was sensitive to the fact that he was not just commenting on Gardener's brain but that, in keeping with her wishes, he was offering the latest contribution to the long-standing question of whether or not there were observable sex differences in human brains.

With regard to this larger question about "sex in brain," Papez was much more tentative and speculative than his nineteenth-century forefathers—just as Gardener would have wanted. Papez included a lengthy subsection on "sex differences in the brain" where he presented a thorough overview of previous research (not including Hammond) and refrained

from making any generalizations about either women's brains or their capabilities. He ultimately concluded that the "chief sex difference is in the
smaller size of the female precuneus" and that the "parietal (precuneal)
index [was] greater in the males."[138] But relative to this finding he wondered if there was a "somatic sex area" in the brain that controlled the sex
organs and that might explain any corresponding structural differences.

If sex had not hampered the development of Gardener's brain, were
there other inherited or environmental factors that might explain its high
development? On this, the answer was clearly yes. In analyzing and explaining the various features of Gardener's brain, Papez often relied on assumptions about race and class and even grouped his comparative brain
samples—forty female and forty male from the Cornell collection—into
"high, middle, and low class" groups. He explained that these classes corresponded with achievement, but they very well might have also correlated
with economic class and race. For example, he reported that Gardener's
"postcentral gyrus" was well developed, as was common "in the brains
of the higher class."[139] Furthermore, Papez drew insight from her race and
genealogy, noting "she has in her ancestry two eminent lines of descent
through Cromwell and Calvert (Baltimore) families. It is evident that a
great mental talent resided in these families who combined the bloods of
the Anglo-Saxon and Celtic races—a talent which was possibly inherent
in her particular mental structure."[140] As a descendent of Lord Baltimore
and Cromwell, of course Gardener's brain would be highly capable.

But Papez should not be considered a biological determinist. Throughout the report he also drew on newer theories about the importance of
environment and culture in shaping one's destiny. While researchers may
have been tempted to look for structural differences in the brain that
would explain social realities, Papez cautioned that "so many conflicting
statements on the association of a more complex degree of cerebral development with intellectual attainment have been made that many recent
authors have been inclined to skepticism. . . . A large number of problems
of cerebral morphology as correlated with anthropology will have to be
solved before useful opinions can be entertained."[141] Papez then stated
matter-of-factly that even though achievement was most certainly a result
of "nervous function," it was also a matter of environment and opportunity: "besides brain function there are so many environmental factors that
have been instrumental in producing men of eminence or renown, it would
seem rash to argue that brain structure alone is an adequate explanation
or that any peculiar wealth of matter is limited to such people."[142] Yet,

even though Papez was careful not to generalize, his report found many more brain differences according to class and race than to sex.

The idea that brains could be classed and raced, but not necessarily sexed, would have appealed to Gardener too, for in many ways that was what she was trying to prove. In lobbying her peers to donate their brains, Gardener, along with Stanton, hoped to align the interests of elite white women with their elite white male peers and distance themselves from poor women, immigrants, and people of color.[143] In disputing Hammond's findings, Gardener questioned the methods by which the brains of female hospital "pickups" had been studied, but she also objected to the fact that brains like hers had not been included. Likewise, she rejected the pseudo-scientific studies of sex differences but also the idea that all women could be grouped in the same category. Race and class, then, played an important role in Gardener's bequest. Gardener did not attempt to overthrow the hi-erarchical ladder of civilization, based upon evolutionary notions of race, that was so frequently invoked by scientists and anthropologists; rather, she wanted to prove that educated, white women had been placed on the wrong rung.

To Gardener's credit, however, she also had the foresight to imagine the complicated ways that nature and culture might interact, as evidenced by her insistence that the brains of infants (rather than adults) be studied for sex differences to rule out the effects of culture, thereby challenging the very foundation of biological determinism. Since the brains of lower animals did not show marked sex differences, Gardener wondered whether any observable differences in human brains might not be "natural and necessary sex differences" but rather "due to difference of opportunity and environment."[144] Therefore, Gardener's resistance to being grouped with female hospital pickups was not necessarily based on her sense that elite white people shared genetic superiority but, rather, that they shared edu-cational and cultural opportunities, opportunities from which women like herself and Stanton had surely benefited.

As Gardener hoped, her brain did what her pen could not: it established once and for all that her intellect had not been handicapped by her sex. Under the headline "Woman's Brain Not Inferior to Men's," the *New York Times* declared that Gardener's brain "posthumously substantiated her life-long contention that, given the same environment, woman's brains are the equal of man's."[145] What a difference fifty years had made. In the 1870s, the *Times* had recommended that all women study Edward Clarke's find-ings about the pathology of menstruation and conduct themselves accordingly, but, by 1925, biological determinism had lost sway and the *Times*, together with *Time* magazine and even *Popular Science Monthly*, forthrightly declared that it was possible for women's brains to equal men's, as if few people had ever thought otherwise.[146] Five years after the ratification of the Nineteenth Amendment, the long battle for suffrage was not paramount in American consciousness when the *New York Times* printed their extensive coverage (five stories in all) of Gardener's brain donation. To *Times* reporters and readers, Gardener was not remarkable as a suffrage strategist, much less as an outspoken freethinker, but as an emblem of the type of educated, professional woman who was quickly becoming commonplace.

With the battle for suffrage won and with Hammond's brain theory discredited by the work of Franklin Mall and others, why did Gardener, then the highest ranking woman ever to serve in the federal government, still feel compelled to donate her brain to Cornell in 1925? After all, nearly forty years had passed since the publication of "Sex in Brain." During this time, Gardener had remarried, traveled the world, played a key role in securing women's right to vote, and served admirably on the U.S. Civil Service Commission. Perhaps Gardener insisted on donating her brain because scientific claims about the biological inferiority of women shaped the thinking of those women's rights activists who came of age in the 1870s and 1880s. Perhaps Gardener was also concerned that twentieth-century women's rights leaders had ceded science to the professional (mostly male) scientists and no longer monitored scientific developments as she and her peers had done in the nineteenth century. While the mainstream press granted ample coverage to Gardener's brain donation, the women's press did not, nor did it cover the pathbreaking experiments discrediting the brain size theory of intelligence or those disproving greater male variability (another Darwinian theory enlisted to prove women's natural inferiority) conducted by social scientists Leta Stetter Hollingworth and Helen Bradford Thompson in the 1910s.[147] In the 1870s and 1880s, women's rights advocates had welcomed science into discussions of sex difference and succeeded in overturning the most biased scientific theories, but their success was predicated on vigilant monitoring of and active participation in science, both of which waned after 1890 as a result of structural changes in the women's rights movement and within the scientific establishment.

One wonders what Gardener thought about the women's movement's decision to cease involvement with science and whether her determination to publicly donate her brain to Cornell was as much a call to action to her women's rights colleagues as it was a final foray into the biology of sex differences. Much like Maria Mitchell, Mary Putnam Jacobi, and Antoinette Brown Blackwell, Gardener knew that women needed science, just as science needed women. Together, these women and their colleagues in the women's rights movement actively shaped the emerging consensus on just how science should be practiced by demanding that scientists structure their studies inclusively, rely on experimental and laboratory evidence (not secondhand anecdotes), interpret their findings objectively, and take seriously the experiences and bodies of women. While she remained hopeful about what objective science might mean for women, Gardener resisted the masculinization of the scientific profession and the contraction of women's rights to focus on the vote. These two transitions shaped the development of twentieth-century science as well as feminist thought, marking the two as mutually incompatible for decades. This is a legacy that Gardener surely would have regretted, for she saw science and feminism as fundamentally allied. Her brain donation represents an alternative to sexist science and unscientific feminism; yet, even though she rejected biological determinism, her bequest also highlights the assumption of white racial superiority that characterized the thinking of many Darwinian feminists and of white Americans more generally. Throughout her lifetime, Gardener witnessed a rash of scientific theories of sex difference come and go, often masking the same conclusions in new studies. She likely suspected that Hammond's theories might one day be resurrected in new garb, as in fact they have been.[148] If so, she understood that her brain might well be women's best defense, both symbolically and literally. As she wrote in her will, "[I]f my brain can be useful to women after I am gone it is at their service through Cornell."[149] One of the ways in which her brain was useful was in its actual study, which substantiated her lifelong contention that women's brains were not hampered by sex. Another way that her brain remains useful is that it reminds those interested in women's rights of the extent to which these rights often hinge on women's bodies, on prevailing definitions of the "natural," and on women's dogged involvement in science. However, as chapter 3 further explores, linking women's rights to women's bodies, and especially to motherhood, has been, historically, a risky strategy as not all women experience motherhood or use their bodies in the same ways.[150] Finally, Gardener's brain bequest illustrates that a central contention of the Darwinian feminists was the right of women, including mothers, to pursue nondomestic tasks and maybe even work outside of the home, another hotly contested debate into which evolutionary science was called as arbiter.

CHAPTER THREE

# Working Women and Animal Mothers

In 1888, Charlotte Perkins Gilman, recently recovered from a traumatic nervous breakdown and the even worse "rest cure" she endured to mitigate it, left her husband, Walter Stetson, and the East Coast behind for a fresh start in Pasadena, near her dear friend Grace Ellery Channing. The years that Gilman, together with her young daughter Katharine, spent in Southern California restored her soul and mind and allowed her to write, lecture, and begin establishing the professional reputation of which she had long dreamed (fig. 3.1).[1] But, six years later, it was her personal life—not her intellectual work—that threatened to permanently etch Gilman's name in the public sphere. In 1894 Gilman received a tempting offer to edit a reformist paper in San Francisco, where she and Katharine were by then living. Gilman had been struggling financially for years and had incurred several debts, making the offer of a full-time job especially appealing. The main problem, as Gilman saw it, was that this arrangement did not seem suitable for a single mother with a young child. So Gilman decided to send nine-year-old Katharine back East to live with her now ex-husband and his new wife, her best friend Grace Channing.

To Gilman, this arrangement made perfect sense, and it seemed perfectly natural, according to her evolutionary philosophy, which challenged the nineteenth century's glorification of self-sacrificing mothers. Moreover, Walter Stetson longed for his daughter, her daughter wanted to know her father (as growing up Gilman had not), and Grace promised to be a much more stable maternal figure than Gilman herself ever could be. Her 1894 divorce, for which Stetson filed on the grounds of "desertion," had caused Gilman's name to "become football for all the papers on the coast," as she recalled in her autobiography; this second unwomanly act was simply too much for the American public to abide. Reporters and editorialists across the coun-

Fig. 3.1. Charlotte Perkins Gilman and her daughter Katharine Beecher Stetson Chamberlin in California, 1893. Photograph courtesy of the Schlesinger Library, Radcliffe Institute, Harvard University.

try pilloried Gilman as an immoral and unnatural woman. What kind of mother would leave her husband and then allow him and his new wife to raise her daughter so that she could pursue a career? Further confounding to the public was that all parties remained on good terms, comfortable, happy even, with this unusual arrangement. As she wrote in her autobiography, "To hear what was said and read what was printed one would think I had handed over a baby in a basket."[2] To mainstream Americans, Gilman had become the public face and ominous future of working mothers.

It was in the context of the resulting self-imposed exile, in the midst of overwhelming public scorn, that Gilman developed the ideas that would characterize her life's work, most notably her landmark *Women and Economics: A Study of the Economic Relation between Men and Women as a Factor in Social Evolution* (1898). This book was Gilman's most important contribution to feminist thought, and it made her an international figure. It also contained, in the words of the historian Carl Degler, "probably the most devastating indictment of traditional nineteenth-century motherhood ever written."[3] Gilman's personal scandal occurred at a time when it seemed that motherhood was on everyone's mind, men and women alike. Not surprisingly, Darwinian evolutionary theory often permeated these discussions, and it informed Gilman's own thinking on what exactly made for, as she called it, "perfect motherhood."

Motherhood was also the subtext, and in many cases the text itself, of discussions regarding the biology of sex difference that animated the women's rights movement in the 1870s and 1880s, as chapters 1 and 2 demonstrated. Motherhood, after all, is what turns abstractions into practicalities in any discussion of the equality or difference of men and women, then as now. In debates about women's intellect and rights, as well as in discussions regarding the proper distribution of domestic and professional work, what people were really talking about was motherhood (actual or potential) and the extent to which it did, or should, determine every aspect of a woman's life. Turn-of-the-century conversations about motherhood hinged on questions of sexual difference, as well as of priority. Whose interests should come first: the mother's or the child's? Furthermore, was motherhood the central, defining feature of a woman's existence? Or one of many roles a woman might play in the course of her life? If, as everyone seemed to agree, motherhood was a priority, how might society help develop the best mothers? To ensure the health of offspring, should one focus on quality or quantity?

Darwin himself did not write much directly about motherhood. He suggested that maternity accounted for the significant differences between women and men and that being mothers was what ultimately limited women's potential to contribute in intellectual or professional realms, as his references to "maternal instincts" in *The Descent of Man* indicated.[4] But in terms of how many babies a woman should have or what a woman might do besides have babies, he did not say. Yet these were precisely the questions that animated public discourse at the turn of the twentieth century. Regardless of Darwin's ambiguity on the particulars, Darwinian language informed all sides of the debate as individuals and groups with varying takes on motherhood sought to couch their positions in evolutionary discourse.[5] Public discussion often centered on fears of "race suicide"—the idea that middle- and upper-class white women were not having enough babies and that soon nonwhite, lower-class populations would become the majority. Concerns regarding race suicide were typically voiced in Darwinian, survival-of-the-fittest language, as previous scholarship has established.[6] But this is only part of the story.

Feminist arguments demanding greater intellectual and professional opportunities for women and mothers, as well as the corollary suggestion that women bear fewer children, also enlisted Darwin. To these reformers, the problem was not that white women were having too few babies but, rather, that the cultural and economic conditions of women's lives were not amenable to either healthy maternity or healthy offspring. Further troubling to Darwinian feminists was the fact that humans were the only species that cordoned off mothers, at least middle- and upper-class white mothers, from productive labor. To many women this seemed not only unnatural but also counterevolutionary because sequestering women in the home hindered the intellectual growth and development of half the species. While many readers may be familiar with Gilman and her take on motherhood, the broader evolutionary context of these turn-of-the-century motherhood debates, especially the important subtext of animal-human kinship, is not as well known. Furthermore, this chapter reveals that Gilman was part of a larger group of women, beginning with Antoinette Brown Blackwell in the 1870s, who were inspired by evolutionary theory and the close relationship between humans and animals that it established to demand that pregnancy be viewed as a healthy, natural process (as opposed to a disease); to suggest that it was better, in evolutionary terms, for individual women and for the species if mothers engaged in the full range of human activities; and to propose a vast rethinking of the home and the workplace to enable mothers to work outside the home because, reformers argued, female domesticity had no precedent among human's relatives in the animal kingdom.[7]

## FIT PREGNANCY

Throughout much of the nineteenth century, middle- and upper-class white motherhood was shrouded in divine, angelic descriptions, providing a core element of the separate-spheres ideology. Advice books and popular magazines such as *Godey's Lady's Book* sought to convince the reading public that the day-to-day trials and tribulations of motherhood—from nursing the sick, to preparing meals, to cleaning bedpans, to wiping away tears—were sacred duties best performed by only the most selfless of mothers who devoted themselves wholeheartedly to the task, even at the risk of death.[8] The first aspect of this sentimental version of motherhood to be challenged by evolutionary science was the long-standing notion that pregnancy was a curse to be endured. By challenging the Adam and Eve myth, evolutionary theory called into question "Eve's curse," the idea that God intended for women to suffer, and maybe even die, during pregnancy and labor to atone for Eve's sin. If there was no Garden of Eden and no original sin, did women still have to suffer Eve's curse? In an evolutionary context, might it not be more important for expectant mothers to think about pregnancy in terms of health, rather than suffering? In this case, traditionalists and progressives seemed to agree that an acceptance of evolutionary cosmology signaled the demise of Eve's curse. New, Darwinian descriptions of motherhood insisted that pregnancy be considered healthy and welcomed as a woman's most important contribution to the evolutionary process. As a result, one idea that Gilded Age reformers of various ideological positions could agree upon was "fit pregnancy."

For most of the nineteenth century, medical and popular opinion had coincided in the belief that pregnancy was, more or less, a disease to be endured. In a chapter entitled "The Pains Attendant on Pregnancy Perhaps Necessary," from a popular 1808 guidebook for women, Samuel K. Jennings advised newly married women that, although symptoms varied from woman to woman, pregnancy to pregnancy, nearly all women could expect their pregnancies to be "troublesome and distressing."[9] Not to worry, reassured Jennings, "the God of nature does nothing in vain," and surely "these distresses seem to be directed to an intended valuable aim." It was not, however, women's business to attempt to figure out what this "valuable aim" might be. The best women could do was endure the pain and hope it bolstered their spiritual strength. Concomitant with the belief that pregnancy was debilitating by design was Jennings's definition of motherhood as primarily a spiritual and moral enterprise, charged with forming the "virtuous affections of the mind."[10] A Darwinian worldview, however, suggested that pregnancy might be better understood in terms of human's animal ancestry, flesh and blood, and hereditary influences, making it harder to sustain the theological justifications for maternal suffering and ignorance.

Many women were keen to this new evolutionary definition of maternity, which, together with Darwin's explanation of human origins, invalidated the Genesis creation story and lifted Eve's curse. Writing in the *Revolution*, Elizabeth Cady Stanton declared that the first step in strengthening young girls was teaching them that pregnancy was not a disability. She suggested that reformers "make all women understand that suffering is not in harmony with God's will." Out of the "ignorance of the science of life" and the pervasive belief that women were the "special object of God's wrath and curse" came "all these absurd theories of the natural weaknesses and disabilities of woman."[11] Stanton realized that the legacy of Eve, especially the curse of maternity, buttressed the ideology of female inferiority. In contrast to biblical explanations of pregnancy, the "science of life" offered women a new lens through which to view maternity and their status as mothers.

Evolutionary theory emboldened many forward-thinking women to denounce literal interpretations of the Bible and, instead, apply the laws of natural science to reproduction. Writing in the *Woman's Tribune* in 1887, E.T. Grover rejected the idea that painful maternity was God's curse upon women for Eve's transgression. Grover declared, "There is a better Eden even in this life, for those who can plant their feet upon the perfect law of life. . . . Then, according to the law of the survival of the fittest, if of no other, we may venture to hope in a better race, and a happier world."[12] Jettisoning the biblical Eden for a Darwinian one, Grover argued that the laws of evolution demanded that mothers be fit and healthy. Or, as Stanton later declared in the *Woman's Bible*, "with obedience to the law of health, diet, dress, and exercise, the period of maternity should be one of added vigor in both body and mind, a perfectly natural operation should not be attended with suffering." Stanton believed that observing "physical and psychical laws" would transform "the supposed curse" into a "blessing."[13]

Numerous turn-of-the-century advice books joined feminists in encouraging healthy pregnancy.[14] In *The Woman Beautiful* (1901), Drs. Monfort Allen and Amelia McGregor noted how the shift from biblical to biological thinking about gender had impacted their ideas on reproduction: "when has the religious world been so distracted by dissensions and differences of opinion? Were there ever as many changes and innovations in theology as at the present time? When did science unfold truths of greater importance and in greater profusion than at this moment?" Since change was so clearly on the horizon, they advocated that it begin in the family. "Let the reform be commenced here, on the principles of physiology and health, and a gradual process of regeneration will be entered on that will produce the most salutatory effects upon the habits, characters, motives, and actions of all mankind."[15] Or, as Dr. Mary Wood-Allen, chair of the Woman's Christian Temperance Union's Purity Department, plainly put it in one of her advice books to women and girls, "I am of the opinion that women were not intended to be invalids in any degree because of their womanhood."[16]

Thanks to advancements in science, popular author Marion Harland noted that the "sacred primal curse" of her grandmother's generation, which they abided and "endured with shame and loathing," had been lifted. She counseled expectant mothers to *"walk regularly, out-of-doors"* because "fresh air and cheerful exercise, the panacea for so many fleshly ills, are never more truly a catholicon than to you, as now situated." To buttress this healthy model of pregnancy, Harland relied on natural analogies: "Pregnancy is no more a disease than is the ripening of a peach, the 'running to seed' of a lily."[17] Because all life had evolved from the same organism, Harland suggested that women could learn lessons about pregnancy even from plants. Eliza Bisbee Duffey's *What Women Should Know: A Woman's Book about Women* (1873) also sought to dispel the myth that women were naturally invalids who should avoid physical exertion at all times and especially during pregnancy.[18] She shared with readers her personal experiences of exercising during pregnancy and her belief that this had resulted in much easier labor, an idea seconded by many others.[19]

Darwinian evolutionary theory provided women, as well as men, with a new vocabulary for understanding themselves as part of the animal and plant kingdoms along with a new appreciation of reproduction as a natural process that followed scientific laws. Of course changing ideas about pregnancy also had much to do with the burgeoning field of obstetrics and the professionalization of medicine, but evolutionary discourse signaled a pivotal point of departure in the turn from religious definitions of pregnancy, marked by suffering and emphasizing the hereafter, to scientific ones, marked by health and stressing the here and now.[20] Over several decades, the naturalistic "fit" model of pregnancy gradually replaced the "disease" model, facilitating larger cultural conversations about motherhood. If women were not permanently disabled by pregnancy, might they also be able to work during and after pregnancies? And, if humans were indeed animals, might it be advisable for human mothers to follow the precedent set by all other female animals and participate more fully in the life of the species?

## "SEX AND WORK"

In answer to such questions, several notable women drew on Darwinian evolutionary theory to argue, yes, it was best for individual women and for the species if mothers pursued intellectual and professional tasks outside the home. After all, such an arrangement would surely improve the health and intellect of women; better accord with the rest of the natural world, with which humans were now an acknowledged part; and likely hasten evolutionary progress by producing healthier babies. Encouraging mothers to pursue professional work would also require that husbands, the home, and the workplace be reconfigured. More than two decades before Charlotte Perkins Gilman published *Women and Economics* (1898), Antoinette Brown Blackwell pioneered the practice of arguing on behalf of working women using evolutionary principles, her own life experiences, and examples from the animal kingdom. For a woman who could boast of numerous achievements and many "firsts," changing public opinion regarding working mothers seemed to be Blackwell's top priority throughout her long life. In this endeavor, Blackwell, found evolutionary science to be an ideal ally.

In histories of the U.S. women's rights movement, Antoinette Brown Blackwell often receives only a brief mention, yet she was a prominent figure in the movement, as well as an important contributor to nineteenth-century feminist thought. In 1850, at the very first national women's rights convention, Blackwell delivered an address refuting biblical injunctions against women speaking in public. In 1920, she cast her vote for President Warren G. Harding, making her the only participant in the 1850 convention who lived to see the ratification of the Nineteenth Amendment.[21] While her formal activities and organizational connections waxed and waned throughout the years, Blackwell was a fixture in the women's rights movement for seventy years. She was also wedded to it by her marriage to Samuel Blackwell, which made her the sister-in-law of Henry Blackwell and his wife, Lucy Stone. Together, Henry Blackwell and Stone led the American Woman Suffrage Association (AWSA) and published the most widely read women's rights newspaper, the *Woman's Journal*, to which Antoinette was a frequent contributor. Despite her close relationship with Stone, she also remained on good terms with the AWSA's one-time rivals, Elizabeth Cady Stanton and Susan B. Anthony, who led the more radical National Woman Suffrage Association and published the *Revolution*.

Blackwell never became a leader in the suffrage movement because she found suffrage activities too limiting. Her two main concerns were establishing the "science of feminine humanity," as described in chapter 2, and changing public opinion regarding working mothers, intellectual interests that mirrored her own personal and professional challenges. As chapter 2 discussed, in 1853, after years of struggle, Blackwell became the nation's first ordained female minister. In 1875, she became the first woman to publish a feminist critique of Darwinian evolution with her book *The Sexes throughout Nature*. In the intervening years, Blackwell abandoned her orthodox faith, left the pulpit, became increasingly interested in science, got married, and gave birth to seven children. Throughout the nineteenth century, she was also a fixture on the reformist lecture circuit, and she published many articles in both the women's rights and scientific presses. If one had to distill a singular message from her long and varied life, it would be this: nature endowed women and men with numerous, complementary gifts. To deny women the opportunity to work and be mothers was both unnatural and unfair. Or as she declared at an Association for the Advancement of Women Congress in 1875, "It is time that we utterly repudiate the pernicious dogma that marriage and a practical life-work are incompatible."[22]

This was a radical position, even for an otherwise iconoclastic woman, to take in the second half of the nineteenth century. Blackwell's papers and unpublished autobiography contain numerous letters from other women's rights leaders debating whether or not it was advisable, or even possible, for women to combine their public work with families. Susan B. Anthony famously chastised women's rights leaders for getting married and having babies because she believed that motherhood must be all-consuming and she feared losing her best allies and workers. After the birth of Blackwell's second daughter, Anthony admonished her: "Now, Nette, not another baby, is my peremptory command. Two will solve the problem whether a woman can be anything more than a wife and mother better than a half dozen or ten."[23] Blackwell's best friend and sister-in-law Lucy Stone admitted that she dared not have more than one baby for fear of becoming overwhelmed herself. After attending a lecture about Joan of Arc and feeling "as though all things were possible," Stone returned home to see her sleeping daughter's face and "shrank like a snail into its shell and saw that for these years I can be only a mother—no trivial thing either."[24] On the other side of the argument sat Elizabeth Cady Stanton, mother of seven, who congratulated Blackwell on the birth of her sixth child, noting, "I would not have one less than seven, in spite of all the abuse that has been heaped upon me for such extravagance."[25] Blackwell ultimately did give birth to seven children; five daughters survived into adulthood (fig. 3.2).

Fig. 3.2. Antoinette Brown Blackwell with her daughter Florence Blackwell (Mayhew),
1876, the year after *The Sexes throughout Nature* was published. Photograph
courtesy of the Schlesinger Library, Radcliffe Institute, Harvard University.

To tease out how and why women might both work and be mothers,
Blackwell delivered numerous speeches on the topic and published two
series of articles in the *Woman's Journal*, "Sex and Work" and "Work in
Relation to the Home," some of which were reprinted in her book *The
Sexes throughout Nature* (1875). Blackwell's writings are an important,
and understudied, nineteenth-century feminist response to evolutionary
theory. Blackwell's interpretation of evolutionary science led her to envi-
sion a world where men and women evolved to become more alike and one

in which the gendered division of labor reflected this change. She shared her feminist interpretation of evolution with her peers in women's organizations, scientific groups, and many journals, and her writings have inspired generations of women who have managed to come across them.[26] Pioneering primatologist and sociobiologist Sarah Blaffer Hrdy, for example, credited Blackwell with being "a true beacon in the night" and lauded her feminist interpretation of Darwin as "the road not taken."[27] Evolutionary theory enabled Blackwell to see humans as part of the animal kingdom (although she believed that human intellectual and emotional faculties distinguished them from other animals and indicated the existence of a divine power),[28] introduced as possibilities the wide variety of domestic and sexual relationships found in the natural world, and convinced her that humans' gendered division of labor was anything but natural.

In her scientific writings, Blackwell did not quibble with the evidence that Darwin, Spencer, and others had compiled from the animal kingdom; rather, she disagreed with the conclusions that male naturalists drew from this evidence and reinterpreted it from a female standpoint, often offering her own experiences in support of her claims. Where Darwin and Spencer saw examples of greater male variability and superiority in nature, Blackwell saw ruling female insects, males who cared for their young, and a tremendous variety of reproductive labor. She listed several animal examples to prove that the more courageous, stronger males praised by Darwin were not necessarily superior to the more highly developed and complex females who could do many unique things such as feed their offspring from their own bodies. For example, she compared the lion to the lioness and argued that while the female is "less strong and valiant in hunting . . . her greater heterogeneity is a full equivalent for this deficiency."[29] Blackwell also praised the male fish who built nests alongside female fish and those who had "the extraordinary habit of hatching eggs laid by the females within their mouths or bronchial cavities."[30] Overall, she rejected Darwin's claim that male superiority was inevitable and cautioned that "the facts of Evolution may have been misinterpreted, by giving undue prominence to such as have been evolved in the male line; and by overlooking equally essential modifications which have arisen in the diverging female line."[31] While she recognized inherent differences between males and females, she suggested that the question of which sex was superior was a matter of interpretation. Darwin believed that male dominance was responsible for human advancement, but, when Blackwell looked to the animal kingdom, she found proof that limiting women's contributions to the domestic realm might in fact have hindered evolutionary progress.

Blackwell pioneered the practice of looking to the animal kingdom for alternative examples of domestic arrangements, a strategy that proved inspirational and rhetorically powerful for nineteenth-century feminists. Comparing people to animals was a useful tool for women because it displaced man as the standard bearer—he became simply one type of organism. Moreover, women's exclusion from productive labor stood out as a peculiarly human construct when compared with labor arrangements in the animal kingdom. As freethinking feminist Helen Hamilton Gardener explained, "Nowhere in all nature is the mere fact of sex—and that the race-producing sex—made a reason for fixed inequality of liberty, of subjugation, of subordination, and of determined inferiority of opportunity in education, in acquirement, in position—in a word, in freedom. Nowhere until we reach, man!"[32] In a later pamphlet, Gardener further insisted that men who demanded complete female devotion to maternity violated the laws of nature. "Nowhere else in nature does the male claim all of the other avenues of life as his special sex privileges," she exhorted, "except alone the one which he cannot perform—that of maternity."[33] When she looked at the animal kingdom to better understand the human condition, Gardener found female equality and autonomy, which starkly contrasted what she observed around her, convincing her of the artificiality of human gender relationships. Gardener delivered a powerful address about motherhood at the 1893 World's Columbian Exposition in Chicago, one of the most high-profile events of the nineteenth century. "Woman as an Annex" attacked the idea advanced by many male evolutionists that men were the "race" while woman was "merely an annex to him."[34] She observed, "Among scientists and evolutionists, and, indeed, even among the various religious explanations of the source and cause of things, the male and female of all species of animals, birds and insects come into life and tread its paths together and as equals."[35] Thus, it seemed obvious to Gardener that, in an age that valued scientific reasoning and evolutionary principles, women needed to be recognized as equal to men and allowed to pursue their own intellectual and professional interests.

Blackwell's analysis of animals, on the other hand, did not convince her that females were equal to males. Rather, she believed that the sexes in all species were different but complementary to each other and that this complementarity enabled each species to survive. To reconcile the seeming contradiction of arguing for equal rights on the basis of difference, Blackwell developed a theory of gender equivalence. In *The Sexes throughout Nature*, Blackwell constructed elaborate tables listing the comparative strengths of each sex throughout the animal and plant kingdoms to illustrate gender equivalence.[36] Blackwell divided all living organisms into two columns, one for male and one for female, and, then, she assigned pluses and minuses in various categories such as size, strength, sexual love, and parental love. Blackwell concluded that the sum total of the male column always equaled that of the female column. In the animal and plant kingdoms, Blackwell found that "greater activity in one sex may fairly balance superior nutritive functions in the other; while, by the law of inheritance, their posterity will be equally advantaged by both, and lifted towards a higher development in both lines of evolution."[37] Men and male animals might engage in a broader range of activities throughout the course of a lifetime, and develop more and varied skills along the way, but all this would be for naught were it not for the females' remarkable ability to provide "direct nutrition" for offspring both in utero and, for mammals, through breast-feeding.[38]

Indeed, it was the female's ability to feed offspring from her own body, both during and after pregnancy, that provided the linchpin in Blackwell's evolutionary schema of sex equivalence and in her calls for the rearrangement of domestic duties. She argued that natural selection demanded the two sexes be equivalent in all realms in order for the offspring to prosper. If the female was responsible for direct nutrition, then the male must provide indirect nutrition. Among terrestrial carnivores, Blackwell observed, the male "forages for the family" while the female provides direct nutrition. Such a system was not practical among herbivores, so "natural selection fixes upon some other division of commensurate duties or acquirements which will be of greatest benefit to the particular species, such as greater beauty of coloring, superior size and strength of muscle, and increased activity of brain."[39] Whereas Darwin had claimed beauty of coloring and superior strength of males as evidence of male superiority, Blackwell emphasized that it was the nutritive function provided by mothers that signaled evolutionary advancement. What enabled humans to develop the cranial and intellectual capacities that separated them from apes, as Blackwell reasoned, was the long gestational period of human babies and the close, hands-on care provided for babies by nursing mothers. Blackwell argued that males and females evolved together so that the males, equally with the females, could "contribute to the general advancement of offspring."[40] Following Blackwell's logic, the evolutionary balance was seriously out of whack in late nineteenth-century America as women provided nearly all the direct and indirect nutrition.

Blackwell was not the only person whose reading of *The Descent of Man* prompted further inquiry into the evolutionary development of sex-differentiated labor. To some, Darwin's suggestion that sex roles had changed over time as organisms evolved was evidence enough of the theory's absurdity. In 1871, the Reverend Wilder Smith of Minnesota observed in a mocking review of *The Descent of Man*, "It is likely that, at this early period, both men and women wore beards. Both also, we are assured, probably bore an equal part in affording innocent lacteal nutriment to their young; an accomplishment some fathers have afforded even within the historical period." Such gender-bending practices as men feeding babies were not limited to the ancestral past; Wilder saw them as part of a dystopian future as well. "It is consoling to obtain here some scientific basis for the women's rights movement," Wilder noted sarcastically. "It is, evidently, a blind and instinctive reversion to a primitive condition in which domestic cares were equally shared by both parents. These then are the applauded results of modern science."[41] For a religious objector like Wilder, evolutionary science portended the dangerous upsetting of tradition, especially with regard to gender roles.

*Popular Science Monthly*, the vehicle through which many Americans learned about evolution, reviewed *The Sexes throughout Nature* in July 1875 and, not surprisingly, took exception to Blackwell's arguments. The reviewer welcomed, albeit in a patronizing way, Blackwell's contribution to the ongoing discussion of evolution but chastised her for attempting to calculate the respective worth of male and female traits through her equivalence tables. To the reviewer, this was an absurd endeavor. Who could devise a system capable of measuring such distinct characteristics? His main objection, however, was that Blackwell did not include maternity, "the grand function of the female sex," as one of women's special functions: "[Maternity] is either left out of the estimate, or must be included under products. Maternity is thus so generalized as to be described in terms applicable to both sexes." The review continued, "Denying, as we do, the equality of the sexes, and holding to the superiority of the female sex, we protest against the degradation of woman implied in losing the supreme and distinctive purpose of her nature among the *plus* and *minus* products common to the sexes."[42] To the editors of *Popular Science Monthly*, what defined women was their total devotion to children, husbands, and housekeeping. Here the editors reiterated the familiar objection to women's rights that maternity somehow set women apart from, perhaps even above, the rest of the species—the very issue that goes to the heart of past and present debates over "difference"—and took Blackwell to task for implying that men should play equivalent roles in child rearing.

Blackwell was guilty as charged. Rather than place motherhood at the top of human activities, she prioritized parenthood. She believed that men should play a larger role in caring for offspring, and she argued that natural selection precluded significant distinctions in both domestic and professional labor.[43] Blackwell's analysis of gender difference was complicated: on the one hand, she firmly believed that sex differentiation indicated evolutionary progress (in agreement with the leading male evolutionists of the day), but, on the other hand, she thought that men and women were evolving to become more alike and that this was positive for the evolution of the species. She suggested that men and women "in search of the same ends" should "co-operate in as many heterogeneous pursuits" as possible.[44] According to Blackwell, the more closely equivalent the sexes and the more similar their daily activities, the more fit and greater in number their offspring would be. To hasten this development, she called for, among other things, greater male involvement in household tasks, full employment and educational opportunities for women, readjustment of domestic power relations, and physical exercise for women.

In a speech before the New England Women's Club, later reprinted in the *Woman's Journal*, Blackwell explained that women working outside the home could not possibly be bad for domestic life. In one passage, she both affirmed the idea that women were naturally drawn to domesticity and discounted the practical importance of this characteristic: "Since love and reverence for the home is ingrained in the feminine constitution, outside work is not likely seriously to impair its interests. Such an influence, if it should ever exist, must be superficial and temporary in its effects; for Nature is not at permanent warfare with herself." Rather than diminish a woman's inherent zest for domesticity, Blackwell claimed that loving one's home life went hand-in-hand with the desire to work outside of it: "The future may do its dressmaking with as much care-saving to the wearer as the present does its tailoring; may banish the washing with the spinning; may order cooked food, as conveniently as the raw material; may send babies to the Kindergartens and young women to colleges, but it may not abolish the home, the family unit."[45] In other words, Blackwell suggested that since some women instinctively felt the urge to work outside the home this desire must be as natural as the desire to work within the home and that both instincts supported the family unit.

Furthermore, Blackwell imagined a future in which men and women devoted an equivalent amount of time to housekeeping and in which men and women alike "should be agriculturists, artisans, scientists, artists, journalists, bookmakers, politicians, statesmen, sea-captains if they will and land-captains, also if they can; heading volunteers in every department of thought and action."[46] Not only did she imagine that women could do masculine jobs, she also imagined that men might do traditionally feminine tasks. At an Association for the Advancement of Women Congress in 1888, Blackwell delivered a speech on women in the professions, noting, "Women of pluck and enterprise can become good ship captains, as successfully as men with pliant fingers can learn to make pretty little things for their wives and daughters to wear; or as men with tender hearts can amuse, instruct, and sympathize with little children." In this same address, Blackwell lamented that women often did not receive equal pay for equal work. To remedy this, she suggested that "if women desire full money values for service, the direct way to gain it is to go into business for themselves." She further recommended that business-owning women "help poorer, less competent women to find better paid openings."[47]

Significantly for Blackwell, women's rights did not just involve the interests of elite women. To create a more equitable society, she proposed that husbands take on larger shares of housework so that lower-class women would not be called upon to do the drudge work of professional women. Blackwell repudiated the "false theory" that "because women are to be the mothers of the race, therefore they are not to be the thinkers or the pioneers in enterprise. This ancient dogma enfeebles one class of women and degrades the other." Blackwell further rejected the idea that "the wife of a laboring man, who accepts ten hours of daily toil as his share of family duty, is bound by her duty to spend twenty-four hours among the pots and the children, with no absolute rest and without fitting recreation." According to Blackwell, such a schema would be bad for men and women of all classes: "If woman's sole responsibility is of the domestic type, one class will be crushed by it, and the other throw it off as a badge of poverty. The poor man's motto, 'Woman's work is never done,' leads inevitably to its antithesis—ladies' work is never begun." Instead, Blackwell proposed "a fairly equal division of duties between men and women."[48] In an 1888 address, she elaborated her position, rejecting the "eight hour husband and the fourteen hour wife."[49] Blackwell's concern for working-class women was based partly on her ideals of social justice and partly on her interpretation of evolution. On the basis of her vast reading in evolution, she believed it would be bad for future offspring if men and women continued to pursue such different day-to-day activities. Although Blackwell herself often employed a servant or governess, she did not write about this or prescribe it for others, as did many other reformist women; rather, she insisted that husbands take on more of the domestic tasks.[50]

To begin with, she suggested that men prepare the food. Throughout the animal kingdom, females provided direct nutrition to the young both before and after birth, while males provided indirect nutrition. Why could men not follow suit? "Let us suppose that natural selection has continually averaged the duty of the sexes to offspring," Blackwell posited. "At maturity, then, males and females would be true equivalents, each equally well fortified to meet its own responsibilities." As Blackwell reasoned, women "should be able to bear and nourish their young children, at a cost of energy equal to the amount expended by [man] as household provider. Beyond this, if human justice is to supplement Nature's provisions, all family duties must be shared equitably, in person or by proxy." Thus, according to the *scientific distribution of work*," men "must be held primarily responsible for the proper *cooking of food*, as for the *production* of it. Since we cannot thrive on the raw materials, like the lower animals, culinary processes must be *allied to indirect nutrition*" (italics in original).[51]

The tax of direct nutrition on mothers was so great that Blackwell further suggested fathers take on additional household tasks to keep nursing mothers comfortable. "It seems to me to be scientifically demonstrable that fathers are equitably bound to contribute indirect sustenance to offspring in the shape of good edible food for the mother," Blackwell reasoned. In addition to cooking "good edible food" for mothers, Blackwell thought that fathers might also be compelled to prepare "ready-made clothing and fires lighted on cold winter mornings" and that if anyone ever had the right to "whine, sulk, or scold . . . because beefsteak and coffee are not prepared for her and exactly to her taste" it was the nursing mother. Furthermore, she exhorted, "if anybody's brain requires to be sacrificed to those two Molochs [a god who requires great personal sacrifice], sewing-machine and cooking-stove, it is not hers!" Nor were Blackwell's plans for domestic rearrangements short term. She proposed that "during the whole childbearing age, at least," a period that could last twenty or more years of a marriage, "if family necessity compels extra hours of toil or care, these must belong to the husband, never to the wife."[52] Even though Blackwell couched her demands in terms of what was best for the offspring, insisting that men cook on a daily basis and take on additional household tasks had the corollary benefit of freeing women to pursue nondomestic work, as Blackwell knew firsthand.

In her writings, Blackwell frequently referenced her own experiences as a college-educated, married mother of five children. To the majority of people who thought that women needed to pick either work or motherhood, Blackwell countered "that [woman] need not sacrifice her own work to any appreciable extent I seem myself to have practically demonstrated.

Any woman with a real work to do in which her whole heart is engaged, in my opinion can do better for the good of the world happily married than if she is single."[53] But, as her biographer Elizabeth Cazden emphasizes, Blackwell's thoughts on work mainly pertained to the sort of work she wished to do (reading, writing, lecturing), not to the needs of women who had to work full time outside of the house, and Blackwell herself switched from actively lecturing to writing from home while her children were growing up.[54] Nevertheless, Blackwell boasted that during her married life she had "averaged certainly three hours of daily habitual brain work, not including daily papers and miscellaneous light reading."[55] While the Blackwells did often employ female servants full or part time, Blackwell gave much of the credit for her exemplary work-life balance to her supportive husband Samuel. When they married in 1856, Antoinette recalled, "it was entirely understood between Mr. Blackwell and myself that my public work would be as nearly uninterrupted as circumstances would allow." Blackwell's husband "declared himself more than willing to help me with home duties," an offer "he generously more than redeemed for almost fifty years."[56] Blackwell's writings exemplified her suggestion that women must "put in evidence the results of her own experience" and provided a template for other educated, middle-class women and men to follow.[57]

Redistributing domestic labor to better accord with evolutionary mandates was the central theme of Blackwell's life and work. Blackwell believed that society was in a state of transition but that eventually "to the majority of women, domestic duties will probably bear about the same relation to outside work as private business does to the majority of men." She also believed that "in all equitable households" men and women should be expected to expend the same amount of time and labor on domestic and professional tasks.[58] Such a system was the only way to guarantee the gender equivalence upon which Blackwell's evolutionary schema rested. To Blackwell, thinking about gender roles in terms of evolution meant comparing humans with animals and rejecting the elements of domesticity she deemed man-made. Increasing men's contributions to household labor would not only ensure gender equivalence but also remove unnecessary, unnatural burdens from women, thereby improving the health of offspring and allowing women greater access to the professions. One reviewer, writing for the San Francisco *Evening Bulletin*, feared that Blackwell might well be on to something, noting "it is easier to pooh-pooh" her argument for greater male involvement in domestic labor "than to disprove its justice." The reviewer concluded that Blackwell's book was a "valuable contribution to the discussion of a great question, and it is the best showing

we have seen from the woman's standpoint."[59] To Blackwell, evolutionary theory provided the ideal justification for women's rights because
it was clear that women's professional labor accorded with natural law:
"Evolution has given and is still giving to woman an increasing complex-
ity of development which cannot find a legitimate field for the exercise
of all its powers within the household. There is a broader, not a higher,
life outside, which she is impelled to enter, taking some share also in its
responsibilities."[60]

## CHARLOTTE PERKINS GILMAN'S REDEFINITION OF MOTHERHOOD

Echoing ideas first introduced by Blackwell in the 1870s, Charlotte Perkins Gilman presented the most thorough case that motherhood, a vital evolutionary activity, need not be the overriding concern of women.
Rather, she proposed that women be encouraged to engage in the full array
of human activities, especially productive labor, in the service of evolutionary advancement. By the time Gilman began actively publishing in
the 1890s, however, the context of women working outside the home had
changed. When Blackwell dared suggest in 1875 that husbands shoulder
more household labor to allow wives to pursue interests outside the home,
the number of white, native-born women engaged in professional work was
very small and included a disproportionate number of women whose last
name was Blackwell.[61] In 1870, the vast majority of women who worked
were employed as servants, and most servants were African American or
immigrant women, not the middle- and upper-class white readers targeted
by Blackwell and Gilman.[62]

By the 1890s, things had changed. Owing in large part to the sex ratio
imbalances resulting from the Civil War, women of all races had entered
college in record numbers and many found new jobs open to them in the
professions. The majority of professional jobs and college degrees were
held by native-born white women, but numbers increased among African
American and immigrant women as well. After 1870, college-educated
white women most often found employment in the newly feminized professions of nursing, teaching, library work, social work, and academics.
By 1890, women made up seventeen percent of the workforce, but thirty-
six percent of professionals were women.[63] And by 1910, twenty percent
of all wage earners were women.[64] At the same time and owing to the
same, postwar sex ratio imbalances, more women remained single, and,
according to historian Rebecca Edwards, the divorce rate "skyrocketed" in

the 1870s and 1880s.[65] With the notable exception of women doctors who were often married, most female professionals were either young college graduates before marriage or women who chose to remain single.[66] Even though most professional women were single, the growing percentages of white women working, together with the steep decline in the birthrate and the rising divorce rate, raised public concern about working wives and mothers.

By the time Gilman began publishing, the most contentious question regarding women's professional aspirations became whether or not such aspirations came at a cost to any future or existing children, echoing concerns raised in the 1870s by Edward Clarke and others about women going to college. In evolutionary terms, was it more important for the mother to be happy, educated, and healthy, or did well-rounded mothers imperil the children? Speaking on behalf of the latter, scientific popularizer Grant Allen published a controversial article in *Popular Science Monthly* in 1889 charging that white women's education and work outside the home had distracted them from their most important duty: procreation. According to "Darwinian principles," Allen claimed that communities must increase in number in order to maintain "national health and vigor." To stem the present threat of declining birthrates among white women, Allen recommended that women marry and have at least four children lest "the race will cease to exist." By "the race," Allen most certainly meant the white, native-born middle and upper classes—those portions of society whose birthrates had dropped the most. Between 1800 and 1900 the birthrate for white American women fell from just over seven to 3.56.[67] He clarified that he supported women's rights but that women's emancipation "must not be of a sort that interferes in any way with this prime natural necessity."[68] To Allen, women were mothers first, humans second.

Allen's proposal that women put off their own intellectual interests and bear at least four children elicited almost as heated a reaction from women as did Edward Clarke's *Sex and Education* sixteen years earlier. In response to Allen, the editors of the *Woman's Standard* pondered, "[M]ust she keep right on reproducing a race of men which turns round and sets its heel upon her neck?" Then, in their "most motherly tones," the editors suggested that "a race born of enslaved mothers is not fit to exist; let it make room for a better one." Furthermore, the editors continued, in order for women to have at least four children, they needed far better prospects for husbands. In the meantime, they advised women to continue with their education and reform work.[69] Responding in *Lucifer the Light-Bearer*, Mary Jordan Finley interpreted evolution to mean that "[r]eproduction no

longer requires the entire energies of one half the race, the world is popu-
lated; henceforth, not increase but maintenance is demanded." As a result,
child rearing was but an "episode in a woman's life instead of the whole of
it, as now" and women could be "released from the overweening predominance of reproductive duties, no longer retar[d] the psychical progress of
the race."[70] Charles Howard Fitch asked "a number of scientific men of the
highest standing" to comment on Grant Allen's suggestion, and he eagerly
wrote the *Woman's Tribune* to report that none of the experts supported
Allen; to the contrary, they informed Fitch that Allen had misinterpreted
Darwin's theories, leading Fitch to conclude "the weight of scientific evidence bears heavily in favor of the enfranchisement of women."[71] While
not writing in direct response to Grant Allen, Dr. Louise Fiske-Bryson conducted her own scientific study to counter a gynecologist's related claim
that educated women lost the desire to be mothers. To "secure some data
upon this subject," she interviewed twenty-five "cultivated women." All
of these women, including three doctors, "have done and are doing sound
solid work that is of value to the world," and all shared the conviction that
"their children are their best creation." As one of the three female doctors told Fiske-Bryson, she "combined from the force of necessity the two
*great* professions—maternity and medicine." Fiske-Bryson concluded that
even though "American women" were often talked about only as mothers,
these "cultivated women" were "also Americans."[72]

Allen's controversial article from 1889 prefigured the highly publicized
race suicide fears that gripped the nation in the first years of the new century and shaped the reception of Charlotte Perkins Gilman's work. At the
turn of the twentieth century, political leaders, together with natural and
social scientists, amplified fears about white women's declining fecundity
by warning of race suicide. Coined by Gilman's friend the sociologist Edward A. Ross in 1901, the term "race suicide" had been a popular idea for
years before it acquired a name. It described white Americans' anxieties
about falling birthrates and the loss of virile manhood as a result of, among
other things, the influx of immigrants to the United States, urbanization,
women's increasing presence in public and professional life, and the loss of
male autonomy in an increasingly corporate world. Women, Ross claimed,
either wasted vital maternal resources on personal endeavors or lost interest in maternity as they pursued education and careers, threatening white
ascendancy.[73]

The American public became familiar with the term "race suicide" in
1903 when President Theodore Roosevelt, the avatar of civilized masculin-

ity, began making it a centerpiece of his speeches and public statements. According to historian Gail Bederman, as soon as Roosevelt mentioned the term it caught the nation's attention. Not only did popular magazines cover the topic exhaustively, but Americans bombarded the White House with letters and photos of large families proudly proclaiming "no race suicide here."[74] In the public imagination, Roosevelt's popularization of the term fused evolutionary progress with white racial superiority and white women's complete devotion to maternity, the latter being the very idea Blackwell and Gilman spent their lives challenging.

Many middle- and upper-class white women shared Roosevelt's concern regarding falling birthrates and encouraged mothers to heed the call to devote themselves to bearing and raising large broods. Roosevelt's rhetoric matched that of the many women's clubs founded on female exceptionalism, including the Woman's Christian Temperance Union (WCTU), which argued that mothers should extend the reach of their nurturing benevolence from their homes out into their towns and cities.[75] Such rhetoric also helped bolster demands that home economics become a recognized academic discipline. Many turn-of-the-century women revivified Edward Clarke's suggestion that female higher education be tailored to prepare women for motherhood, not jobs outside the home. Rather than seek outside careers or interests, these women encouraged their peers to turn motherhood itself into a profession, complete with ample tools and a specialized education. Such demands often included evolutionary language about the survival of the species and encouraged women to take their maternity seriously, for the sake of their own offspring and the progress of humanity in general.

The National Congress of Mothers (NCM), which became the Parent-Teacher Association (PTA) in 1924, best exemplified women's interest in this more conservative take on evolutionary motherhood. Founded in 1897 by two elite clubwomen, Alice McLellan Birney and Phoebe Apperson Hearst (mother of William Randolph Hearst), the NCM modeled itself on the General Federation of Women's Clubs but focused on educated motherhood, not issues of concern to professional women. According to Christine Woyshner's study of the group, the NCM had three priorities: parent education, child welfare, and home-school relations. The NCM's focus and federated organizational strategy matched the Progressive ideology of the time period and mirrored that of many other women's clubs. Their emphasis on maternity also aligned the group with the race-suicide popularizers including Roosevelt and Allen. The NCM did not support suffrage and blamed the higher education of women for encouraging women to abandon their traditional domestic pursuits, leaving them unprepared for the realities of maternal life.[76]

Charlotte Perkins Gilman challenged, in word and deed, the maternalistic rhetoric of organizations like the NCM. In stark contrast to conventional wisdom, she suggested that those women who developed themselves as individuals made the best mothers. In *Women and Economics* (1898), Gilman drew heavily upon her interpretation of Darwin and other evolutionists to argue that the main problem with society was that women were economically dependent upon men, a fate without parallel in the animal kingdom. Paraphrasing Darwin in *The Descent of Man*, Gilman explained that in the state of nature, natural selection served as a check against excessive sexual distinction. The peacock's tail could never become too bright or too big, for example, because a gigantic tail would impede the individual's survival. Among humans, however, Gilman contended that natural selection now promoted excessive sexual distinctions because women depended on men for food.[77] In other words, for women, men *were* the environment, a situation that did not tend to foster the strongest, healthiest women. She hoped her work would help remove the economic element of marriage by allowing women to become financially self-sufficient and, thus, restore the proper balance between natural and sexual selection. Like Blackwell, Gilman called for a revolution in public and private life to facilitate women's entry into the paid labor force.

The impact of *Women and Economics* matched Gilman's powerful demands. The book went through seven editions, was translated into seven languages, and was frequently used as a college textbook in the years prior to 1920. The *Nation* declared *Women and Economics* "the most significant utterance on the subject [of women] since [John Stuart] Mill's *Subjection of Woman*." And Carrie Chapman Catt, the president of the National American Woman Suffrage Association, placed Gilman at the top of her list of the dozen greatest American women.[78] Readers also noted the connection between *Women and Economics* and *The Descent of Man*. Writing in the *Labor Advocate*, Ella Ormsby proclaimed that "there is a new book which ought to be read in cool, quiet places." She then compared Gilman's work to Darwin's, noting, "Out of Darwin's Descent of Man which tore down and rebuilt the foundations of the world's philosophy one idea was distorted and mockingly passed about by busy little orators who hadn't had time to read the book. . . . something like this should be said of this new work on sociology."[79] Building on Darwin's reframing of human-animal kinship and on popular social evolutionary theories promising progress toward perfection, Gilman sought to redefine the relationship between men and women. In fact, she began *Women and Economics* by explaining that evolutionary theory necessitated a rethinking of human domestic relations: "Since we have learned to study the development of human life as we study the evolution of species throughout the animal kingdom, some peculiar phenomena which have puzzled the philosopher and moralist for so long, begin to show themselves in a new light."[80] This new light was the light of the "natural," and, to Gilman, it simply was not natural for women, at least middle- and upper-class white women, to be corseted, weak, and removed from meaningful labor. Lower class and non-white women did not often have the luxury of being removed from paid labor, but they too tended to be dependent upon men for survival (even if they also worked) and they too had their labor segregated by sex.

Gilman's life experiences also informed her passionate opposition to women's economic reliance on men. She chafed at being financially dependent on her first husband, struggled financially for several years after leaving him, and witnessed firsthand the economic challenges married women faced while she was growing up in a household with an absent father.[81] In her autobiography, Gilman recalled that "more than once I saw my mother without any money or any definite prospect of any."[82] After a string of suitors and broken engagements, including a fiancé who died suddenly of typhus, Gilman's mother Mary married her father Frederick Beecher Perkins at the relatively old age of twenty-nine. Unfortunately, Frederick Beecher Perkins did not turn out to be the marrying kind and soon abandoned the family. So Mary and her two small children (two others died in infancy) were shuffled from relative to relative, neighbor to neighbor, for much of Gilman's youth. In her autobiography, Gilman movingly described her mother's plight:

> Mother's life was one of the most painfully thwarted I have ever known. After her flood of lovers, she became a deserted wife. The most passionately domestic of home-worshipping housewives, she was forced to move nineteen times in eighteen years, fourteen of them from one city to another. After a long and thorough musical education, developing unusual talent, she sold her piano to pay the butcher's bill and never owned another.[83]

This fear of economic uncertainty and dependence upon men shaped Gilman's worldview and, no doubt, inspired her to put female economic self-sufficiency at the very heart of her reformist agenda.

To Gilman, women's economic dependence on men was not just a personal problem for individual women; she believed it also stalled evolutionary progress for the "race." Whether she meant the "human race" or the "white race" has been hotly debated by historians. Led by Gail Bederman, some scholars argue that because Gilman internalized the racial hierarchy of civilization implicit in popular evolutionary discourse she did not need to specify that she meant the white race because it would have been obvious to her readers.[84] Because of her emphasis on the "discourse of civilization," moreover, Bederman claims that Gilman's feminism was "at its very base racist."[85] In contrast, other historians, notably Gilman biographer Judith Allen, argue that Gilman meant the "human race" and that Gilman's ideas regarding race were nuanced and progressive for her time. For example, Allen points out that Gilman rejected as "class legislation" the arguments for "educated suffrage" that many of her fellow feminists supported, as well as actively opposed the virulent racism of her day including the lynching, disenfranchisement, and segregation of African Americans.[86] While turn-of-the-century ideas about race surely informed Gilman's understanding of gender and while she likely imagined women much like herself when crafting her arguments, as Bederman so persuasively argues, Allen presents much evidence to support her claim that, at least most of the time, Gilman meant the "human race" when she wrote the "race" and that racism is not the defining characteristic of her vast writings.

Key to Gilman's understanding of evolution was sociologist Lester Frank Ward's "gynaecocentric theory," first presented in an 1888 article in the *Forum* and later elaborated on in *Pure Sociology* (1903) and other works.[87] Ward was the first president of the American Sociological Association, a longtime federal government scientist, and one of the era's "most brilliant and unconventional thinkers."[88] He is best known for popularizing reform Darwinism, an evolutionary sociological system that starkly contrasted that of Herbert Spencer and the American social Darwinists, led by William Graham Sumner, though it did share their assumption of a racialized hierarchy of civilization. Whereas social Darwinists lobbied to abandon the poor and let only the fittest survive, Ward suggested that what made humans "human" was their ability to care for one another, change their environment, and, thus, shape the future.[89] As part of the reformed future that Ward imagined, women would play a much larger role in society, including in the labor force. Ward believed that it was natural for women to be on equal footing with men in all realms of society and that cultural constructs, specifically religion and patriarchal marriage, had kept women from their rightful place for much of recorded history. Taking the long view of history made possible by evolutionary and geological discovery revealed another possible prehistory to Ward, a past in which women and mothers reigned supreme.

By the 1880s, according to Ward biographer Clifford Scott, Ward had become disenchanted with the political focus of women's rights efforts and "saw greater opportunity for the emancipation of women through science education and outside work."[90] In assessing recent scientific findings regarding sex difference, Ward was surprised that "those who start out avowedly from a Darwinian standpoint should so quickly abandon it and proceed to argue from pre-Darwinian premises."[91] How could a theory based on reproduction deny the principal reproducers the lead role? And, if scientists had in fact established human-animal kinship, why should women's lives differ so drastically from men's when such gendered distinctions did not characterize other animal species? In answer to these questions, Ward posited that "androcentrism"—by which he meant male supremacy—was a relatively new phenomenon in evolutionary terms, following eons of rule by women. Ward argued that evolutionary science demonstrated beyond a doubt that "woman is the grandest fact in nature" and that "the elevation of woman is the only sure road to the evolution of man."[92] A cornerstone of the future egalitarian society imagined by Ward was women working alongside men, participating in leisure activities along with men, and wearing simple, comfortable clothes like men.[93] Ward also suggested that humans needed to return to female choice of sexual mates to better accord with evolutionary mandates and the rest of the animal kingdom, as will be discussed in chapter 4. Ward's influence on Darwinian feminists, especially Gilman, was foundational because he supplied them with a unified theory linking evolutionary progress with feminist goals, authored by the highly credible "father of American sociology."[94]

With "reverent love and gratitude," Gilman dedicated *The Man-Made World or Our Androcentric Culture* (1911) to Ward because "all women are especially bound in honor and gratitude for his Gynaecocentric Theory of Life, than which nothing so important to humanity has been advanced since the Theory of Evolution, and nothing so important to women has ever been given to the world."[95] Evolutionary theory unlocked the mystery of creation and established human-animal kinship; Ward added to that a new appreciation for women and maternity. In *The Man-Made World*, Gilman charged that society's emphasis on masculinity and femininity had caused people to overlook our "common humanity," a situation without precedent in the animal kingdom. As a remedy, Gilman redefined "man's work"—which she described as "every handicraft, every profession, every science, every art, all normal amusements and recreations, all government, education, religion"—as "human activities" that, by definition, women should be able to participate in as well.[96]

Of course, Gilman was hardly a strict Darwinian, or an exclusive follower of Ward's for that matter; she also drew heavily on Edward Bellamy, Herbert Spencer, and Jean-Baptiste Lamarck. And it is important to note that the subtitle she chose for *Women and Economics* was *A Study of the Economic Relation between Men and Women as a Factor in Social Evolution*, indicating her interest in sociological, as well as biological, theories of evolution and her belief that evolution meant progress. Her varied reading in evolution led her to believe that improving the social environment would benefit both current and future generations. Like Ward, Gilman can best be understood as a "reform Darwinist," someone who thought that people could shape the evolutionary process and who believed that evolution was inherently a progressive force, leading to ever more development and opportunity for all.

Although not an orthodox Darwinian, her worldview was profoundly shaped by the Darwinian revolution and her "conversion" to evolution. In fact, Gilman considered evolution to be her religion. In her autobiography, Gilman described her search for a meaningful truth system, one that answered the basic question "why are we here" and also passed intellectual muster. Ultimately Gilman decided that "social evolution," the application of evolutionary laws to society, was her religion: "I figured it out that the business of mankind was to carry out the evolution of the human race, according to the laws of nature, adding the conscious direction, the telic force, proper to our kind—we are the only creatures that can assist evolution."[97] While Gilman drew heavily on Spencer as well as Darwin, it was Darwinian evolutionary theory that introduced her to rigorous scientific study and encouraged her to probe the concept of the "natural;" it also provided her with the language to critique female domesticity and the pivotal idea that lessons about human society could be drawn from the animal kingdom.

Nor was Gilman solely interested in proposing a solution to female economic dependence on men; she was also interested in determining how this type of marriage came to predominate in the first place. Precisely when and why had human society become androcentric? In addition to drawing on Ward's anthropological account of the origins of patriarchy, she placed a large part of the blame on Christianity and its explanations of creation and gender difference. In 1912, Gilman revealed that, like the women chronicled in chapter 1, she too was initially drawn to evolutionary theory because it refuted the story of Adam and Eve. As Gilman explained, "Our ideas are all based on the primal concept expressed in the Adam and Eve story—that he was made first, and that she was made to assist him. On this assumption rests all our social structure as it concerns the sexes." If people could "reverse this idea once and for all . . . all our dark and tangled problems of unhappiness, sin and disease, as between men and women, are cleared."[98] Gilman recognized that the legend of Eve shaped popular thinking about gender and hindered the women's movement, and she hoped that a reformist interpretation of evolution would free women from the legacy of Eve.

So central was the Garden of Eden story to Gilman's understanding of female subjugation that she began *Women and Economics* with her own creation story in the form of a "proem."[99] In Gilman's version of Eden, "twofold man was equal" until man found the "Tree of Knowledge" and realized that he could rule over woman, his former comrade, by keeping her weak. Men and women had continued in this state of inequality until the end of the nineteenth century, when, Gilman argued, it was no longer in their evolutionary best interest to do so. Gilman assured readers that her reevaluation of gender roles would not lead to free love, anarchy, or the demise of the family. To the contrary, she suggested that allowing women to become economic producers would make them better wives and better mothers and that divesting marriage of its economic function and freeing the mother from her servant duties would make for stronger families as well as healthier children.

According to Gilman, Americans had become inured to excessive distinctions between men and women, unnatural differences that thwarted evolutionary advancement, because they had been taught to believe they were necessary for the idealized version of motherhood that characterized nineteenth-century thought. Popular reverence for motherhood discouraged people from thinking critically about it. Gilman exhorted her readers to "turn the light of science and the honest labor of thought upon this phase of human life as upon any other." After all, "motherhood is but a process of life, and open to study as all processes of life are open."[100] If mothers were to be accorded such lofty praise and excused from paid labor on account of maternity, shouldn't scientists at least determine whether or not nonproductive women made the best mothers, Gilman asked. According to her interpretation of evolutionary science, human mothers paled in comparison to their animal counterparts, who raised far more healthy offspring with far fewer resources: "The human mother does less for her young, both absolutely and proportionately, than any kind of mother on earth." Gilman concluded there was no "special superiority in human maternity."[101]

At the root of this undeserved glorification of human motherhood was the tendency to cloak it in divine, rather than naturalistic, terms. Instead of teaching young women about the physiological and psychological demands of motherhood, Gilman charged that Americans presumed motherhood would be "fulfilled by the mysterious working of what we call 'the divine instinct of maternity.'" Gilman countered that "maternal instinct is a very respectable and useful instinct common to most animals. It is 'divine' and 'holy' only as all the laws of nature are divine and holy; and it is such only when it works to the right fulfillment of its use."[102] Using evolutionary principles and animal examples, Gilman challenged the nineteenth-century glorification of the mother, the main building block in the development of separate-spheres ideology. The nonproductive (outside the home, that is), highly feminized mother typified nineteenth-century depictions of women as well as many women's estimates of themselves. Not only did Gilman refuse to emulate the idealized mother in her own life, she had the temerity to suggest that such women did not in fact make the best mothers.

Gilman's cross-species analysis of maternal practices—itself an evolutionary endeavor—in *Women and Economics* opened her eyes to new domestic arrangements and to what she believed were the strengths of animal mothers. When Gilman looked to the animal kingdom, she saw examples of harmonious distribution of labor that suggested to her that human housework should be reconfigured to make it cooperative and remunerative, for the sake of both mother and child. Distributing household tasks, such as cooking and cleaning, to paid specialists would increase the value of the work, enable women to become economically productive members of society, and eradicate marriages based on economic necessity. As a first step, Gilman proposed replacing individual kitchens in homes and apartments with centrally located ones where trained professionals would prepare the food: "Eating is an individual function. Cooking is a social function." According to Gilman, neither were "family functions."[103] As later critics have pointed out, however, Gilman assumed that women—paid or otherwise—would still be the ones performing domestic labor and, although she did not specify, it is likely that lower-class women would end up doing the domestic work of middle-class, professional women. In contrast to Antoinette Brown Blackwell who enlisted the help of supportive husbands, Gilman, who did not marry a supportive husband until later in life, proposed a system whereby some women would turn domesticity into a profession so that others, presumably those with more resources, could pursue more interesting careers.

Gilman's ideal domestic model was a feminist apartment hotel, in which individual women, together with their nuclear families if they had them, could live in private apartments but utilize professional cooking, cleaning, and child care services located in the same apartment complex. Gilman's disciple Henrietta Rodman even raised money and enlisted an architect to have one such apartment hotel built in New York City; however, the plan did not materialize. By relying on lower-class, less-educated women to do the housekeeping and child rearing so that better educated women could pursue professional work, Gilman limited her support base among the class of women most likely to be working. At the same time, the number of women included in Gilman's ideal target audience of professional working mothers was very limited because most female professionals were single.[104]

Not surprisingly, Gilman's premise that women's emancipation was predicated on professionalizing domestic labor to allow women to work outside the home and to better accord with evolutionary mandates generated much discussion among women themselves. While many left-leaning women lauded her work, others were not convinced. The New England Women's Club inadvertently debated *Women and Economics* during a group discussion of whether or not women should work for money if they were not forced to do so by necessity. Apropos of this question, Mrs. White brought up *Women and Economics* and declared its influence as "misleading and pernicious."[105] The *Clubwoman*, the organ of the General Federation of Women's Clubs, critically reviewed Gilman's *Concerning Children* (1901), which elaborated on ideas first enunciated in *Women and Economics*. The reviewer recognized that it was a brilliant book "however much the average mother may quarrel with her judgment or reject her conclusions." The reviewer also took offense to Gilman's suggestion that there might have been a better way to raise a family: "Mrs. Gilman is inclined, we think, to go to extremes, especially when she takes the ground that the average, well-educated and affectionate mother is wholly unfit to bring up her own children."[106] Clearly, Gilman was treading on sacred terrain.

Gilman's most strident opponents, however, were women with their own competing interpretations of evolutionary motherhood. Elizabeth Sloan Chesser's *Woman, Marriage, and Motherhood* (1913), which was endorsed by the NCM, relied on evolutionary principles to argue that women who worked only in the home were the best sort of mothers (although

Chesser herself was a doctor). Chesser traced the evolution of maternity from the single cell to the birds to the mammals. Along the way she found much to praise, including the male stickleback who "builds the nest for the young, safeguards his wife and offspring, and is an excellent help-mate—an example to many irresponsible husbands much higher in the life scale."[107] But she distinguished between maternity, which humans and animals both experienced, and motherhood, which was the sole province of women because it encompassed "ethical" and "psychological" attributes. According to Chesser, previous civilizations treasured these maternal virtues during the earlier "mother age," but the advent of Christendom, especially the Reformation, doomed women to their current second-class status. While Chesser agreed with Gilman that "biologically, the mother is paramount" and "socially, the mother is the basis of racial progress," she believed that women needed more sex-specific training, not less. To ensure that mothers would not have to work outside the home, Chesser adopted the mainstream evolutionist stance to argue that "it is the dif-ferences between the two sexes that provide the most valuable evolution-ary factors." Relying on the same basic texts as Gilman, Chesser came up with the exact opposite solution. She critiqued *Women and Econom-ics* and its "amaternal theory" of professionalized housekeeping and child care for being "against nature, biology, the lessons of human evolution."[108] Instead, Chesser suggested that the government set up a Home Depart-ment, state hospitals for mothers, insurance for mothers, and pensions for mothers. Rather than dismantle the private home, Chesser proposed that it be institutionalized.

Contemporary critics, as well as later scholars, frequently juxtaposed Gilman's evolutionary vision of motherhood with that of influential Swedish feminist Ellen Key who was also an enthusiastic Darwinian.[109] In her intellectual biography of Gilman, Judith Allen maintains that, in fact, Gilman refused to be called a "feminist" in her lifetime (she preferred the term "humanist") largely because the term "feminist" was so often asso-ciated with Ellen Key.[110] Key, too, played up these differences by frequently citing Gilman's "amaternal" work as the antithesis of her own.[111] In *The Century of the Child* (1909), Key laid out her proposal for the future of motherhood and the family. Based on her belief in evolution and her inter-est in the latest hereditary and eugenic theories, Key argued that women should devote themselves solely to maternity. Moreover, Key suggested that all parental and societal decisions be based on the best interests of the child, not the mother. In an interesting twist, however, she included more liberal views toward divorce and extramarital sexual relations under the category of things that might benefit children. Children prospered in happy homes, she wrote, and people should celebrate sexual and romantic relationships, wherever they are found, in order to maximize happiness.

Key referred to evolution as "the holiness of generation," and in her worldview it replaced religion as the ordering principle of life, as it had for Gilman. She argued that people should make decisions based on the future growth and development of the race, not on what may or may not happen in the afterlife. She believed that "the greatest obstacle to the free discussion of this theme [the relations between the sexes] is still the Christian way of looking at the origin and nature of man."[112] For too long, women had been taught to model themselves on the Virgin Mother and view sex as shameful. Key rejected this view and proposed that "we must on the basis of natural science attain, in a newer and nobler form, the whole antique love for bodily strength and beauty, the whole antique reverence for the divine character of the continuation of the race, combined with the whole modern consciousness of the soulful happiness of ideal love."[113] Christianity taught that the body was sinful, whereas evolution taught Key that the flesh deserved worship. Rather than study the scriptures, Key exhorted men and women to "learn the laws of natural selection and act in the spirit of these laws."[114] Unlike Gilman, Key demanded the total dedication of women to maternal functions. She believed that each individual mother was the person best suited to meet every one of her child's needs—educationally, emotionally, nutritionally, and health-wise—and that women could never equal men in the professions as a result of this all-encompassing responsibility. "I have shown more than once that woman by her maternal functions, uses up so much physical and psychical energy," observed Key, that in the "sphere of intellectual production she must remain of less significance."[115] Edward Clarke could not have said it better himself.

Gilman rejected Key's demands for what she called "primitive motherhood," just as an earlier generation of women rejected Clarke's similar propositions. She believed that household and child-rearing duties were better performed by trained specialists because doing so would increase the level of job performance as well as liberate individual housewives to follow their own professional interests and develop themselves as individuals. According to Gilman, "a mother who is something more—who is also a social servant—is a nobler being for a child to love and follow than a mother who is nothing more—except a home servant." Children, including even babies, would learn more and develop better if they spent part of the day taught by trained experts in an educational environment rather than in a "small isolated building, consecrated as a restaurant and dormitory for one family."[116]

Gilman also criticized Key for thinking of women as females first, humans second. To Gilman, this seemed utterly unscientific and opposed to evolutionary progress. As she wrote in the *Forerunner*, "Ellen Key, with the rest of the world, fails to recognize that distinction of species is far larger and more important than the distinction of sex. Our humanness is a quality common to both sexes."[117] In later iterations, Gilman took this a step further, arguing that women were in fact superior to men.[118] To her, it seemed absurd to sequester mothers, the most highly developed individuals, in the private home and force them to do the drudge work of the species.

Neither Key nor Gilman doubted that society needed to be reformulated in order to ensure healthier offspring and evolutionary progress, but they disagreed about how individual women and mothers could contribute to this revolution. To Gilman, well-developed women, who were humans first and mothers second, offered the most to their children. To Key, women's humanity hinged on motherhood. Whether they thought of women first as humans or as mothers, however, both Gilman and Key agreed that society needed to reorganize the home, relations between men and women, and child rearing to foster evolutionary progress. Antoinette Brown Blackwell was perhaps the most forward thinking in this regard by insisting that husbands and marital customs needed to change to allow working mothers greater professional opportunities.

The debates over whether or not women and mothers should work outside the home highlighted the various ways in which women interpreted and applied evolutionary discourse to questions of gender and sex. Beginning with Antoinette Brown Blackwell's suggestion that men should take charge of the cooking, women concerned about motherhood applied evolutionary theory to numerous, often contrasting, ends. But these sometimes contradictory applications also demonstrated the new synthesis of ideas about motherhood facilitated by Darwinian evolution, namely, an emphasis on its physicality, a new appreciation for the body, and more forceful demands for female control of reproduction. Blackwell and Gilman, together with Key, were all enthusiastic converts to Darwinian evolution, which powerfully reshaped their ideas about motherhood and reproduction, and all redefined or rejected outright Christian definitions of maternity and the family. Each in her own way also promulgated new ideas about the body—Blackwell suggested that women use their bodily experiences and expertise to develop the "science of feminine humanity"; Gilman thought women should focus on bodily health, not beauty, as did Key, who added to that an appreciation for women's nonreproductive sex drives. Both Gilman and Key also focused on the physicality of maternity and advocated franker discussion of reproduction as well as greater acceptance of sex education.

Central to these various suggestions regarding motherhood and domesticity was the profound rethinking of women's role in the reproductive process necessitated by the acceptance of Darwinian evolution. Blackwell, Gilman, and Key contrasted evolutionary models of marital roles with biblical ones, and each appreciated that evolution invalidated the cursed pregnancy and wifely subordination prescribed by Genesis. By allowing women to imagine an evolutionary cosmology in which humans were closely related to animals and where health—not salvation—of offspring was the most important goal, Darwinian evolutionary discourse allowed women to reimagine the relationships between husband and wife, mother and children, leading to demands for fit pregnancy, the equitable distribution of domestic labor, and, for some, the entrance of women and mothers into the paid workforce. Darwinian evolutionary theory also led many women and social progressives to demand greater female control of reproduction more generally, as chapter 4 will chronicle, because this, too, accorded with animal precedent and, thus, seemed only natural.

## CHAPTER FOUR

# "Female Choice" and the Reproductive Autonomy of Women

In 1885, Midwestern suffragist and socialist Eliza Burt Gamble (1841–1920) spent a year in Washington, D.C., probing the vast collections of the Library of Congress (fig. 4.1). Gamble had become convinced that women were superior to men, despite the scientific and popular consensus to the contrary, and she sought academic evidence on which to ground her conviction. Like the other Darwinian feminists profiled in this book, she had initially thought about gender relations in terms of the Bible, but she had come to believe that organized Christianity was built upon the subjugation of women. In her later years, she turned her attention to science and socialism. Raised by a widowed mother and orphaned as a teenager, Gamble supported herself as a teacher in Michigan, which perhaps influenced her lifelong commitment to women's economic self-sufficiency. In 1865, at the age of twenty-three, she married James Gamble, a successful lawyer and businessman who supported her feminist and socialist activities financially and intellectually. The Gambles had three children, one of whom died in infancy; their two surviving children, Helen and William, were both given the combined last name "Burt Gamble," an unusual decision testifying to Gamble's lifelong interest in reforming dominant reproductive patterns.[1] As Gamble was starting her own family, she simultaneously embarked on a career as a writer and reformer. She attended the very first Michigan State Suffrage Society meeting on January 20, 1870, and also served on the group's executive committee.[2] In 1880, she was inspired by an antisuffrage speech to publish her own critique of the gospel.[3] By 1882, she had reached the conclusion that "the female organism is in no wise inferior to that of the male."[4] In 1884, she gave a "very able" speech on "Woman and the Church" at the annual Women's Rights Convention in Minnesota, where she and her family briefly lived before returning to

Fig. 4.1. Eliza Burt Gamble, in *The National Cyclopaedia of American Biography: Being the History of the United States, Volume 18* (New York: James T. White & Company, 1922), 220–1. Reproduced from the Collection of the Library of Congress.

Michigan.[5] But, after fifteen years of activism and contemplation, Gamble still sought an organizing belief system that could support her feminist principles. While reading in the Library of Congress in 1885, she found one: Charles Darwin's theory of sexual selection, as explained in *The Descent of Man, and Selection in Relation to Sex* (1871).

After carefully studying the *Descent*, Gamble "became impressed with the belief that the theory of evolution, as enunciated by scientists, furnishes much evidence going to show that the female among all orders of life, man included, represents a higher stage of development than the male."[6] She believed that male scientists had overlooked or misinterpreted what the theory of evolution meant for questions of sex difference. In particular, she was surprised to find that Darwin and other scientists had amassed all the evidence for female superiority and yet "seemed inclined to ignore certain facts connected with this theory which tend to prove the superiority of the female organism."[7] To share her insights regarding what evolutionary theory meant for women, Gamble published a book entitled *The Evolution of Woman* (1894), later revised as *The Sexes in Science and History: An Inquiry into the Dogma of Woman's Inferiority to Man* (1916). According to Gamble, the evidence for female superiority, as well as the answer to the nation's most pressing social problems, was the Darwinian concept of "female choice" of sexual partners, which Darwin explained was the norm among all animals except for humans.

Over the years, Gamble increasingly combined her feminism with a commitment to socialism. Her attention to the economic aspects of women's subjugation furthered her interest in female choice because she believed that financial independence was a precondition of free choice. Feminist socialists, most notably Gamble and Charlotte Perkins Gilman, seized on female choice because it linked their objections to capitalism with their concerns about patriarchy. These reformers believed that capitalism and patriarchy worked together to ensure that women remained subordinate to men, especially through marriage customs. Because most women could not support themselves financially, they had to marry a man, often any man, in order to survive, throwing off the natural process of sexual selection by introducing money into the equation. Thus, female choice presented feminist socialists with one unified way to critique the institution of marriage, decry the lack of economic opportunities for women, denounce capitalism for creating a class of wealthy men for whom fitness was not a criterion to mating or success, and reject the type of women—corseted, frail, and submissive—so often selected as wives by men.

To reformers, sexual selection also provided reassuring evidence that humans could intervene in the evolutionary process through their reproductive decisions. As emerging scientific research discredited Lamarckian explanations for the heritability of acquired traits (although many people continued to cling to this idea), female choice provided a new way for reformers to think about intentional, intergenerational change. Unlike natural selection in which a cold, unknowing environment gradually weeded out weak traits, sexual selection suggested that individuals could help shape the future of their species, consciously or otherwise, through their mating choices. This especially appealed to American interpretations of evolution, which fused faith in science with a commitment to continual improvement and progress. Voicing their demands for change in the language of female choice allowed reformers to claim that their ideas were not radical but natural since, according to Darwin, female choice governed mating customs in all species except *Homo sapiens*.[8] In contrast, woman's subordinate, ornamental status in patriarchal, capitalistic societies stood out as unnatural because it had no precedent in the animal kingdom. Moreover, after Darwin spent the bulk of *The Descent of Man* arguing for human-animal kinship on all levels, his denial of female choice among humans alone seemed arbitrary to feminist reformers, if not to naturalists. If it was natural and desirable for females throughout the animal kingdom to select their mates, and if humans were indeed animals—linked to other species in both mind and body—why was female choice not the norm among people?

At the turn of the twentieth century, female choice reverberated widely throughout feminist and socialist reform circles in the United States and abroad. Reformers did not seem concerned, or in some cases even aware, that most naturalists rejected the theory of sexual selection and the predominance of female choice in animals.[9] For these reformers, it was enough that Darwin, the great naturalist himself, had articulated the concept and given it his imprimatur, thereby granting scientific credibility to their cause. Furthermore, feminist reformers were not so much concerned with whether or not female choice occurred among animals but, rather, whether it *could* occur among humans. A return to female choice among humans not only promised to cure many modern social problems, it also provided a naturalistic rationale for the type of social progress that feminist socialists envisioned. Eventually, this line of thinking shaped the ideology of the feminist socialist who initiated the broad-based campaign for birth control in America: Margaret Sanger.

## FEMALE CHOICE IN *THE DESCENT OF MAN*

As Darwin described in *The Descent of Man*, female choice of mates shaped the evolutionary process throughout the animal kingdom, including among primitive man, by determining which traits would be passed on to the next generation. In a Darwinian world, "eager" males competed with each other for access to females, while females were "coy" and selected the most ornamented and vigorous males.[10] For female choice to be effective, most, if not all, of the females had to select for the same, or similar, traits, lest the selected traits cancel each other out. As Darwin reasoned, female choice had resulted in the breathtaking plumage of peacocks, the mind-boggling variety of horns and antlers, and the marvelous uniqueness of bird songs. Throughout *The Descent of Man*, Darwin described female choice as the motive force creating the peculiarities of each species as well as the natural beauty found among all living things.

Darwin presented considerable evidence to demonstrate that female choice was the norm among birds, and, based on bird customs and the observations of numerous animal breeders, he mused "it would be a strange anomaly if female quadrupeds, which stand higher in the scale and have higher mental powers, did not generally, or at least often, exert some choice."[11] Among humans, however, Darwin posited that women had lost the power of selection during the more "savage" years of human history. Primitive females had selected men for strength and vigor (tendencies that fostered the transition from ape to human); in turn, the resulting strong men had wrested the power of selection away from women. For their part, men tended to pick only the most beautiful women as mates with little regard for their overall fitness or health (except to the extent that beauty indicated health). In addition to ingrained Victorian ideas about the differences between men and women, Darwin was led to this conclusion by his observation that, among humans, women were the more highly ornamented sex, a telltale marker of membership in the category "selected," not "selector." In modern times, women attracted and men selected (except in rare instances among "utterly barbarous" tribes where female selection still took place).[12] Loss of female choice, as Darwin described it, appeared to be a requisite change in the transition to becoming fully human.

While many elements of sexual selection theory struck Darwin's fellow naturalists as dubious, few objected to his hypothesis that high-status men selected the most beautiful women as mates, leaving other men to select among the remainder. This seemed obvious and beyond scrutiny. What many naturalists did object to, however, were the twinned ideas that animals could exercise choice and that human romance was simply one form of animal mating. Naturalists recoiled from these suggestions because they blurred the animal-human boundary and removed any notion that humans were "specially created," precisely Darwin's points in *The Descent of Man*. To eliminate the need for an omniscient creator, Darwin had to offer an entirely naturalistic explanation for all aspects of organic life, including humans' mental and emotional capacities, which he claimed differed only in degree, not kind, from those of animals.[13] Human-animal kinship on this level was simply too much for most of Darwin's colleagues to accept. It was one thing to acknowledge that humans and gorillas shared a common ancestor but quite another to imagine that they also shared thoughts and feelings.

Alfred Russel Wallace, the codiscoverer of natural selection, led the charge against sexual selection. Wallace was deeply invested in maintaining the mystery of human uniqueness and, with it, the need for a divine creator. Wallace and Darwin exchanged numerous letters on sexual selection, and their disagreements only increased over time. With each ultimately unable to convince the other, Wallace published a full-scale critique of sexual selection in his book *Darwinism* (1889). To Wallace, the traits and habits Darwin ascribed to sexual selection—colorful plumage, bird song, huge antlers—could more accurately be explained by natural selection, a mechanism linked to survival, not intentional choice. For example, he thought that antlers and other weapons of attack were the result of natural selection (because the stronger males would be more likely to survive as well as mate), and he thought the comparatively dull coloration of most female birds was likely a protective mechanism, adapted to camouflage mothers guarding eggs. In rejecting sexual selection, Wallace claimed to be more Darwinian (meaning adhering to natural selection) than Darwin, hence Wallace's decision to title his own book *Darwinism*.[14]

Above all, Wallace objected to Darwin's assertion that animals could demonstrate a semblance, even the most primitive one, of rational choice or aesthetic discernment. Whereas Darwin stressed the permeability of the animal-human boundary and offered a naturalistic explanation for every human mental and emotional state, Wallace believed that only supernatural causes could have created the human mind and that this rational mind, in turn, fundamentally separated humans from animals. Lacking the ability to rationally distinguish between suitors or appreciate beauty, how could a female animal "choose" a mate?[15] Indeed, according to Erika Lorraine Milam's comprehensive study of scientific engagement with female choice, many nineteenth- and early twentieth-century scientists shared Wallace's concerns and dismissed the idea that animals exercised choice-based behavior.[16]

Of course, many of the scientists who rejected sexual selection theory also had trouble embracing the idea that females, animal or human, could have exercised such agency. As Darwin himself observed in 1882, most naturalists "admit that the weapons of male animals are the result of sexual selection," but "many naturalists doubt, or deny, that female animals ever exert any choice, so as to select certain males in preference to others."[17] For his part, Wallace noted that Darwin's suggestion that "all the ornaments and colours of birds and insects" had been "produced by the perceptions and choice of the female, has . . . staggered many evolutionists." And Wallace hoped his colleagues would feel "relief," as he himself had, when his objections to sexual selection seemingly eliminated the need for female choice by explaining extraordinary male features "as dependent on the general laws of development, and on the action of 'natural selection.'"[18]

Other scientists were less politic when rejecting female choice. St. George Mivart, a long-standing and outspoken foe of Darwin and natural selection, framed his objection to sexual selection theory around the obvious absurdity of female choice, noting that "such is the instability of a vicious feminine caprice, that no constancy of coloration could be produced by its selective action."[19] On the one hand, scientific critics rejected female choice at least partly because the concept offended their ideas about human distinctiveness and the proper role of women, but, on the other hand, as these strident rejections reveal, female choice offered other readers the possibility of feminist interpretations of sexual selection.[20]

## SURVIVAL OF THE FRILLIEST?

To understand the folly of male choice of sexual mates, many feminist reformers claimed one needed only look at women's fashions. To be considered beautiful at the turn of the twentieth century, a woman was typically corseted, high-heeled, and frail. How could a woman who could scarcely walk a mile produce strong offspring, reformers wondered? If one was concerned about healthy offspring, should not women be selected for being fit, strong, and intelligent? Sociologist and reform Darwinist Lester Frank Ward regretted that the eons of men selecting women based on their beauty had resulted in the loss of "the greater part of all those sterling qualities that primarily characterize the female sex as the original trunk of all organic existence and the source and prop of life itself."[21] Female fashions demonstrated beyond a shadow of doubt that men were not capable of selecting mates based on fitness or health and foreshadowed all too vividly the sort of offspring that would result from unions of strong, lascivious men and dainty, ornamental women. Moreover, it did not go unnoticed that many of the most fashionable turn-of-the-century women donned bright feathers and other icons of male animal finery that Darwin had described in such detail in *The Descent of Man*, presenting a visual confirmation that selection had indeed been turned upside down.[22] Signaling their acceptance of the eager male/coy female binary, reformers argued that, unlike men, women were not governed by lustful desires and would, instead, rationally select mates based on health, fitness, and probity. Thus, a return to female choice would be doubly effective: not only would it encourage women to be healthy rather than decorative, it would also serve as a check on male sexual desire.

In *The Sexes in Science and History* (the revised version of *The Evolution of Woman*), Eliza Burt Gamble analyzed the merits of male versus female choice among humans and animals. Among animals, Gamble argued that male traits developed through sexual selection were not "in the line of progress" or "true development."[23] With regard to male aesthetic flourishes like tusks and bright feathers, Gamble reasoned that the "female has made the male beautiful that she might endure his caresses."[24] In contrast to the arguments of Darwin and other male evolutionists that bright feathers and strong tusks established the superiority of the males, Gamble concluded that such traits were merely wasted energy, representing costly diversions from other more useful functions. "Secondary sexual characters do not assist their possessor in overcoming the unfavourable conditions of his environment," Gamble reasoned, "they are not within the line of true development, but, on the contrary, as their growth requires a great expenditure of vital force, and, as is the case among birds, they often hinder the free use of the legs in running and walking, and entirely destroy the use of the wings for flight, they must be detrimental to the entire structure."[25] Females, on the other hand, used their vital force for more useful functions, such as pregnancy and nursing, further establishing the superiority of the female.

Among humans, Gamble argued that before women had lost the power of selection, they had used it to select for, and in essence create, the type of man that enabled the advent of civilization: "While the female has been performing the higher functions in the processes of reproduction, through her force of will, or through her power of choice, she has also been the directing and controlling agency in the development of those characters in the male through which, when the human species was reached, he was enabled to attain a limited degree of progress." This also indicated to Gamble that women were superior to men, for the "stream may not rise higher than its source."[26] If women's choices had enabled humans to become humans, weren't women due the credit for the advance of civilization?

Whereas female choice of mates had fostered the development of both humanity and civilization, male choice, according to Gamble, had stunted evolution by sacrificing the health of women and future generations to the vagaries of men's shortsighted and lusty whims. Although women certainly appreciated beauty, female aesthetic sensibilities could not possibly explain their choice of dress: "on the contrary, it is to Sexual Selection that we must look for an explanation of the incongruities and absurdities presented by the so-called female fashions of the past and present." She argued that woman, whose "business in life has been to marry . . . in order to gain her support," has been forced "by her charms [to] captivate the male." Women had no choice but to wear uncomfortable, injurious

clothing simply to appeal to male taste in the hopes of attracting a husband. As Gamble caustically observed, "The girl at the ball with the wasp
waist and the greater number of furbelows [ruffles] is never a wall-flower
and her numbers never go unfilled."[27] Corsets and long skirts prohibited
free movement and natural growth, yet women still wore them, despite
the risks. Here, Gamble appeared to accept the idea that men were naturally more lustful than women, but she also boldly critiqued the sexual
objectification of women. To Gamble, the most obvious way to counter the
sexual objectification of women was a return to female selection of mates.

Central to Gamble's intellectual mission was identifying the precise
historical moment at which women had lost the power of selection. According to her scientific, theological, historical, and anthropological research, Gamble concluded that the advent of wife capture—a practice that
she believed emerged as humans fully evolved from their ape ancestors—
signaled the moment when women no longer selected for men. In searching for a prehistoric time before patriarchy, Gamble was part of a much
larger nineteenth-century intellectual tradition. According to Cynthia
Eller's study of matriarchal prehistory, its popularity can be traced to the
1861 publication of Johann Jakob Bachofen's *Mother Right: A Study of the
Religious and Juridical Nature of Gynecocracy in the Ancient World*. Bachofen's research especially influenced the developing field of anthropology, which took the idea of a matriarchal prehistory as a given. American
proponents of matriarchal prehistory were also inspired by the work of
Lewis Henry Morgan, especially his *Ancient Society* (1878), and by Frederick Engels's *The Origin of the Family, Private Property, and the State*
(1884). Engels located the "world historical defeat of women" with the advent of private property and the resulting privatization of domestic work.[28]
To him, a matriarchal golden age preceded private property and, through a
Marxist revolution, could be attained once more. As a committed socialist
and amateur anthropologist, Gamble would have been familiar with the
ideas of Bachofen, Morgan, and Engels, and her own descriptions of the
matriarchate were surely informed by theirs. Gamble should also be considered in the context of the nineteenth-century cohort of women, most
notably Matilda Joslyn Gage, who wrote nostalgically about a lost mother
age. What distinguished Gamble's prehistory from that of Engels, Morgan,
or Gage, however, was her desire to ground her matriarchate in Darwinian
evolutionary tenets, specifically female choice.

Among humans, Gamble charged that women had been victimized by
male choice and the resulting "excessive and useless maternity" for too
long.[29] Gamble believed that in addition to selecting decorative mates, men

chose to have sex frequently with little regard for the health of women or the toll that pregnancy took. Like many turn-of-the-century women, Gamble may well have been voicing a very logical fear of frequent pregnancies and maternal death. Throughout the nineteenth century, the maternal death rate in the United States remained comparatively high even as the birthrate fell. At the turn of the twentieth century, it hovered at seven deaths per one thousand live births, to say nothing of the babies who died at birth or within their first year.[30] Gamble lost one of her three children in infancy, and her own mother passed away when Gamble was just fifteen. Gamble, no doubt, knew countless other women who suffered even greater losses in childbirth and was, thus, voicing a personal, realistic concern about the link between female reproductive autonomy and the overall health of women and children.

Gamble's discovery of sexual selection theory in 1885 heightened her commitment to the cause of women's sexual autonomy, which became a key theme in her writing. In the 1900s and 1910s, Gamble wrote several articles in reform publications as well as letters to the editor of her hometown newspaper, the *Detroit Free Press*. She returned again and again to the topic of women's reproductive agency and the sexual double standard. Often, she joined other socialists in countering the prevailing fears regarding race suicide outlined in chapter 3. To socialists, race suicide was, essentially, a capitalist propaganda campaign designed to produce the maximum number of workers (though Gamble did not point out that the middle-class white women who were the target of race suicide propaganda were not, generally, the women giving birth to factory workers). As she wrote in the *International Socialist Review*, President Theodore Roosevelt used to speak in favor of woman's suffrage but "since he has become powerful and has taken upon himself the responsibility of maintaining the capitalist regime, nothing has been heard from him relative to the self government of women. His only advice to them is 'bear children.'"[31] Gamble objected to what she called "enforced maternity" on two grounds: first, she believed women alone should decide whether or not to have children; and, second, she favored quality, not quantity, when it came to offspring because she believed that fewer wanted babies would be better for the evolutionary process. "Under higher human conditions" and armed with economic independence, Gamble argued, women's desire for offspring and their own good sense would "regulate the birth rate."[32] On these grounds she even opposed motherhood pensions, a fairly popular idea among reformers, because she thought such a program would "commercialize" reproduction, just as sex itself had been commercialized, and encourage women to have children

when they were not in fact ready. Rather than institute government programs to regulate the birthrate, Gamble again suggested that women look
to evolutionary precedents in the animal kingdom: "Everybody who has
observed the conditions and habits of the species below man knows that
civilized (?) woman, as far as her reproductive functions are concerned, oc-
cupies a position much inferior to that of the female animal."[33] The main
thing distinguishing animal reproduction from human in Gamble's mind
was female choice of sexual partners.

Taken together, Gamble's written opus advanced several related points.
First and foremost, she hoped to establish the evolutionary superiority of
women and female animals. A closely related goal was to critique church
teachings on the inevitability of female inferiority based on Eve's curse.
In the 1910s, Gamble wrote several letters to the *Detroit Free Press* on the
topic of prostitution. She was very concerned lest reform efforts perpetuate the sexual double standard by arresting, prosecuting, and medically
examining only the female prostitutes and not their male patrons. She believed that the answer to the "vice problem" was to first recognize that its
root cause was the fundamental hypocrisy of the Christian church and its
conflicting messages about women as sexual temptresses or virgin mothers: "The institutions which foster and uphold vice can no longer conceal
their underlying principles. Scientific investigation, which has revealed
and is still revealing many facts relative to mans' origin and development
is slowly, but surely undermining the ignorance which is responsible for
the long continued degradation of women." Once society recognized this
mistaken biblical origins story and replaced it with an evolutionary one
that restored women's "queenship" in sexual selection, then, Gamble
imagined, in only three generations "free women" would succeed in eradi-
cating history's oldest profession.[34]

Finally, Gamble strove to establish that women's reproductive autonomy and economic independence were key preconditions to any meaningful
reform. As she pointedly observed, "since women as economic and sexual
slaves have become dependent upon men for their support, no male biped
has been too stupid, too ugly, or too vicious to take to himself a mate and
perpetuate his imperfections. This unchecked freedom of the male to multiply his defects is responsible for the present conditions."[35] Such statements resonated with the critiques of marriage as "legalized prostitution"
that were made by Elizabeth Cady Stanton and free love advocates (a category that did not include Stanton) throughout the nineteenth century
and were bolstered by Gamble's familiarity with sexual selection theory.[36]
While Gamble appears to have married for love and enjoyed a happy mar-

riage, her teenage years as a self-sufficient orphan surely impressed upon her how hard it was for single women to support themselves. Gamble's writings highlight the faith that socialist reformers had in female choice, as practiced by economically independent women, to simultaneously rid the world of prostitution and vice, elevate the status of women, rein in lustful men, and ensure that future children would be well cared for.

While female readers may have applauded Gamble's arguments in *The Evolution of Woman* (1894), *The Sexes in Science and History* (1916), and her many articles, several male readers in the mainstream press expressed outrage at her temerity and her conclusions. *The Evolution of Woman* was advertised and reviewed in newspapers and magazines across the nation, from the *New York Times* to the *Chicago Tribune* to the *San Francisco Chronicle*, indicating the broad appeal of its scientific theme. Indeed, the sheer number and geographic distribution of the reviews of *The Evolution of Woman* establish that Gamble, while little known today, was widely read in her time.[37] Most reviewers praised the scholarship of the book, but several dismissed Gamble's conclusions as "biased" and, worse, as fomenting "sex antagonism."[38] Even the legendary journalist H.L. Mencken attacked Gamble in his book *In Defense of Women* (1922), which was not in fact written in defense of women. Mencken fundamentally misread her argument as supposing that female choice still predominated among humans and dismissed such a conclusion as absurd. He also included several ad hominen attacks on Gamble in his four-page screed, describing her as a "gyneophile theorist with no experience of the world."[39]

Summarizing its read of the work, the *New York Times* titled its review "A Fearless Assault on Men." "Under a scientific garb, this book treats an old question from a somewhat original point of view," observed the *Times*. The reviewer conceded that arguments for women's equality were not new, but "we have not heard, however, that woman is, by the nature of her organization, the extent of her development, and her primary characteristics, man's superior."[40] Characteristic of many of the reviews of Gamble's work, the *Times* also scolded Gamble for inserting ideology into science, precisely what she (and many other women) accused male naturalists of doing. Echoing the mainstream reviews of other feminist forays into science described in chapter 2, the *Times* cautioned, "It should be remembered that true science, which implies a willingness to seek long and patiently after truth, seldom is vituperative." Nevertheless, the *Times* review concluded on a sympathetic note: "In spite, however, of the harshness, crudity, and the, perhaps, necessarily disagreeable line of thought, the book contains much that is true, many suggestive facts cleverly compiled, and a spirit of indignation that is not unnatural when we contemplate the centuries during which women were victims to the lower motive forces of man."[41]

That the crux of Gamble's argument was really about female sexual autonomy (both the right to choose one's sexual partner and to determine the frequency of intercourse) was evident in the *Popular Science Monthly* review of *The Evolution of Woman*. This reviewer noted that Gamble's sections on history were "much stronger" than her scientific sections and criticized her for not scientifically establishing her claims regarding female superiority, showing again how contested the definition of science was and how it often depended as much upon the gender of the investigator as the nature of his or her conclusions. Regarding Gamble's claim that maternal love was more prevalent among animals than paternal love, the reviewer made a particularly nasty attack: "hyperexaltation of [parental love] often follows thwarting or lack of sexual love which is its natural antecedent."[42] In other words, according to this reviewer, Gamble's problem was that she was frigid, and her lack of a healthy sex life accounted for her general grumpiness and overzealous critique of men.

*The Evolution of Woman* was not universally panned, however. In addition to being favorably recommended in women's rights and freethought publications, it received positive reviews in the *Nation*, the *Critic*, and *Current Literature*, which praised the work for its "great research made in a truth-loving spirit."[43] Here again, "truth" and "science" seemed to be in the eye of the beholder. Foreshadowing more modern interpretations of Gamble's work, the *Critic* praised her for establishing how "the dogma of [female] inferiority, has been taught from generation to generation, until, like the little girl with the doll, it is accepted without inquiry or thought." This review also considered Gamble's work to be scientific, noting that it was "based on Darwin's 'Descent of Man' and the works of other noted scientific writers . . . [and] it is not a harangue on Women's Rights, but a careful and scientific treatise." The *Critic* appreciated that the book was "written out of a full heart and by a person who has something to say and knows how to say it" and concluded that it was "worth buying and reading."[44] Similarly, the *Nation* recommended the book to general readers, noting that "whoever enjoys an admirable piece of argument, set forth in an admirably lucid and convincing manner, will take pleasure in Mrs. Gamble's book."[45] Gamble's book also inspired at least one other woman to publish her own volume tracing the history of women's degradation, laying the blame on the mistaken biblical idea that men were created first and seeing hope for the future in Darwinian evolution and female choice of sexual mates. In the acknowledgments to her 1912 book *The Advance of Woman: From the Earliest Times to the Present*, author Jane Johnstone Christie thanked "Mr. Darwin, Mr. [Lewis Henry] Morgan, Mr. Lester F. Ward, Eliza Burt Gamble, [and] John Stuart Mill."[46] Impressive company for a schoolteacher from Michigan. Taken together, these various reviews of Gamble's *The Evolution of Woman* establish just how contested, and gendered, the definition of science was in the late nineteenth century and how appealing the concept of female choice was to feminist and socialist reformers, if not to the general public.

## THE SOCIALIST ARGUMENT FOR "FEMALE CHOICE"

At the turn of the twentieth century, the Darwinian concept of female choice of sexual partners gained traction among socialists. Throughout the Gilded Age, many reformers expressed concern that industrial capitalism did not reward the "best" people and, worse, that the selfish tendencies fostered by capitalism would be passed on to succeeding generations. Rather than reward decent, hard-working people, capitalism allowed the unscrupulous ones to prosper and leave their fortunes to their undeserving offspring. Moreover, reformers feared that the excesses of capitalism, together with lax inheritance laws, had resulted in a situation in which inordinately rich men had unfair advantages in mating. Regardless of how attractive, kind, or smart a wealthy man was, he could nearly always select a beautiful young woman to marry and, then, proceed to reproduce as much or as little as he pleased. While rich men snatched the most desirable mates, the remaining men—who may have had better personalities, keener intellects, and stronger bodies—often did not prosper under capitalism and, thus, were not considered the top picks when it came to marriage. Compounding this problem was the fact that in modern society women's marital choices were hampered by the economics of patriarchy. Since most women could not support themselves financially, many were compelled to marry men whom they wouldn't necessarily have chosen simply to have a roof over their heads and food on the table. In a socialist society, women would not have to depend upon men for food and could then select mates based on attraction and affection.

So strong was the appeal of female choice to socialists that by 1890 Alfred Russel Wallace, the leading scientific opponent of sexual selection, began advocating female choice among humans, the very concept he had spent years trying to discredit. After applauding the recent educational and professional advances made by women, Wallace suggested that the

driving factor in any meaningful societal reform would be female choice
of marriage partners: "I hope I make it clear that women must be free to
marry or not to marry before there can be true natural selection in the
most important relationship of life . . . In order to cleanse society of the unfit [and allow natural selection to proceed]," he explained, "we must give
to woman the power of selection in marriage, and the means by which this
most important and desirable end can be attained will be brought about
by giving her such training and education as shall render her economically independent."[47] Wallace still did not think female choice occurred
among animals, which he believed were incapable of exercising choice.
But, among rational humans, he suggested that female choice could be a
conscious, political attempt to thwart the counterevolutionary tendencies
of capitalism.

Wallace's position on female choice also depended on his acceptance
of eugenic ideas about the "weeding out" of the "unfit," as evidenced in
his statement above, but he rejected state-sponsored eugenics in favor of
empowering individual women to make the best choices for themselves.
Unlike organized eugenicists who campaigned for state, medical, and legal
limits on marriage, Wallace and other feminist socialists believed future
offspring would simply be better born when economically independent
women could make reproductive decisions for themselves. In September
of 1913, the *Masses*, a popular socialist magazine in the United States,
cheered Wallace for uniting evolution with women's rights:

> Sex selection and the survival of the fittest are held mainly responsible
> for the course of evolution. Sex selection means the choice—especially
> by the female—of superior mates. Hence the elimination of inferior
> qualities in posterity. This free act of natural passion is what has lifted
> and conserved the race. Alfred Russel Wallace—known as the co-
> discoverer with Darwin of these principles—says that our civilization,
> in making women economically dependent upon men has destroyed
> the action of the first principle. The lifting and conserving power of
> woman's choice is lost, because her choice is not free. For there is no
> freedom but that is grounded in economic independence.[48]

Here the *Masses'* editors reiterated Wallace's main point that it was unnatural and counterevolutionary for one sex to be dependent upon the
other for survival, showing the appeal of female choice among feminist
socialists in the United States. Moreover, the *Masses'* praise for Wallace

demonstrated that left-leaning reformers felt affirmed by enlisting Wallace and evolutionary science more broadly to their cause.

What prompted Wallace's historic and highly publicized change of heart regarding female choice was reading Edward Bellamy's utopian novel *Looking Backward 2000–1887* (1888), one of the best-selling books of the nineteenth century.[49] In fact, *Looking Backward* became only the second American novel (after *Uncle Tom's Cabin*) to sell one million copies.[50] Translated into several languages, it also inspired a mass political movement in America, known as "Nationalism." In the 1892 presidential election, the Populist candidate General James B. Weaver adopted several Nationalist principles and garnered a million votes.[51] After the book's publication, "Bellamy clubs" sprung up across the country and Bellamy became a national figure until his death from tuberculosis in 1898.[52]

Wallace biographer Martin Fichman describes Bellamy as Wallace's "mentor" and details at length the transformative experience Wallace had as a result of reading *Looking Backward* soon after it was published. The working-class Wallace had long been drawn to socialist ideas but feared they were too authoritarian; likewise, he knew that human reproductive choices directed the future evolution of the species, but he rejected state-sponsored controls on marriage as similarly authoritarian and also too elitist. In Bellamy's socialist utopia, Wallace found precisely the balance of individual freedom and social equality for which he had long searched. As Fichman explains, Wallace thought Bellamy's brand of socialism would remove the "disparities of wealth and rank" and "eliminate the economic and political prejudices that . . . dominated the selection of reproductive partners in Victorian society. In their place, mate choice would focus on those higher moral and intellectual traits often neglected (or rendered subservient) in competitive capitalist society."[53] Wallace elaborated on his socialist ideas in several works published between 1890 and his death in 1913, beginning with his essay "Human Selection" in which he declared himself a socialist as a result of having read *Looking Backward*. Years later, as he surveyed his life's accomplishments, Wallace, the codiscoverer of natural selection, concluded that female choice by women in a socialist society was "by far the most important of the new ideas I have given to the world."[54] Although, as Fichman makes clear, Wallace largely borrowed this idea from Bellamy.

In the futuristic world Bellamy created in *Looking Backward*, female choice and economic independence were the two interdependent engines of socialist utopia. Because women and men earned the same wages in

Bellamy's ideal society, women could freely choose partners based on attraction or remain single if no suitable partners could be found. In a world where money was not a precondition to marriage, Bellamy felt convinced that women would select only the wisest, kindest, and most healthy mates. According to Bellamy biographer Sylvia Bowman, women's lack of professional options and dependence on marriage for economic survival had long been a concern of his; he was an early supporter of women's suffrage and wrote about the perils of marriage by economic necessity as early as 1873.[55] Similarly, Sylvia Strauss notes that Bellamy was strongly influenced by his good friend Thomas Wentworth Higginson, the Civil War hero, abolitionist, and prominent women's rights activist. Like several of the most progressive women's rights activists, Bellamy repeatedly cited marriage practices as especially detrimental not only to women but also to society.[56] As Bowman explains, "Bellamy was opposed to loveless marriages that were made for economic or social reasons because they interfered with the working of Darwin's theory of natural selection and because women, if they were not to be a torment to themselves and others, should marry those they loved—and so should men if they sought sympathy, companionship and helpfulness from their wives."[57] Although he did not advocate state regulation of marriage (he believed education and social pressure would compel individual women to select only the best possible mates for themselves), "Bellamy expected Darwin's theory of natural or sexual selection, which he regarded as essential to the betterment of humanity, to influence the 'physical, intellectual and moral character of the race.'"[58] In 1878, Bellamy wrote notes for a story about "breeding superior souls" and then published "A Love Story Reversed," which portrayed "an ideal society in which marriages were arranged according to the theory of natural selection."[59] These ideas came to fruition in *Looking Backward*. The overwhelming popularity of *Looking Backward* introduced readers around the world to the Darwinian concept of female choice, suggested that female choice could play a major role in social progress, and linked the concept to socialist reform.

## FEMALE CHOICE IN CHARLOTTE PERKINS GILMAN'S UTOPIA

Bellamy's *Looking Backward* also inspired Charlotte Perkins Gilman, perhaps the best known of the Darwinian feminists. When Gilman moved to California to start a new life, she joined a Nationalist club, published her very first article in a Nationalist periodical, and became one of the movement's most popular speakers. During her time with the club, she developed the ideas that she later elaborated on in *Women and Economics* (1898), namely, the connections between women's lack of economic and marital choices.[60] Gilman's ideas about female choice echoed those of Gamble, though there is no evidence that the two knew each other or read each other's work. Gilman's and Gamble's critiques of male choice focused on the frequency of male demands for sex (constant as opposed to periodic) and the type of partners men selected (weak and pretty, not strong and healthy). The two women independently came to the same conclusion, that if women could support themselves economically they would select better partners and wisely limit reproduction, and everyone would be better off. Both Gamble and Gilman further believed that female inequality was cultural not natural, that women at one point reigned supreme, and that women's subjugation happened when they were forced to enter into marriage for survival. Darwin's suggestion that, at an earlier evolutionary stage, women had lost the power of selection allowed them to date the beginning of female subordination and tie it to both sex and economics. To remedy the situation, Gilman, too, called for the resumption of female choice.

In *Women and Economics* (1898), published just four years after Gamble's *The Evolution of Woman*, Gilman launched the sustained critique of male choice that characterized much of her life's work. According to Gilman, women had become too feminine in their attempts to attract mates; this "over-sexing" of women thwarted not only individual development but also evolutionary progress. To correct this, society would have to rethink the mores governing marriage, the family, and the home. Furthermore, this situation hindered the evolutionary process because men selected women according to their "femaleness," not according to their overall "fitness" as humans.

Both Gilman and Gamble drew on sociologist Lester Frank Ward's "gynaecocentric" theory, introduced in chapter 3, which also promoted a return to female choice. Throughout many eons of human existence on earth, Ward contended that women had controlled cultural and sexual relations, only to be supplanted when men—who had been made larger and stronger than women as a result of generations of women selecting for size and strength—realized how reproduction worked and began to suppress women in order to guarantee paternity. Ward's gynaecocentric theory of history sought to establish that "in the economy of nature the female sex is the primary, and the male a secondary element. If this be a law, its application to the human race is readily made and its importance to social life

cannot be ignored."[61] Gamble and Gilman agreed with Ward about the revolutionary potential of female choice and did much to popularize his work
among other feminists. Indeed, a 1917 critique of feminism laid much of
the blame for the movement's success at the feet of Ward, "the founder of
the prevalent naturalising school of feminists."[62] The author elaborated:
"This idea of sexual selection *by the females* makes a strong appeal to the
feminists; and it may be said that Darwin was the originator of modern
feminism, and Ward is his prophet."[63]

In linking economics with marriage, Gilman etched out a distinctive
difference between her version of gynaecocentrism and Ward's.[64] Not only
did Gilman reject the assumption that she borrowed Ward's theory (she
believed their influence on each other was symbiotic), she also disagreed
with Ward's version of how women lost the power of selection. As Judith
Allen establishes in her intellectual biography of Gilman, Ward believed
that men had taken away the power of selection through violence and rape,
whereas Gilman focused on the economic elements of women's subservience and their limited job options.[65] To Gilman, the economic aspect of
marriage, not so much the sexual, was the fundamental problem revealed
by male choice of mates and, in turn, an essential part of the solution.
Throughout her life Gilman identified with various strains of socialist
thought—first Nationalism, then Fabianism (the British school of socialism known for advocating gradual change not strikes or demonstrations).
By the 1910s, Gilman felt increasingly at odds with the masculinist socialist movement in the United States, but she still maintained faith in the
redemptive power of collectivist socialistic goals. As Allen observes, "the
entirety of [Gilman's] feminist analyses presupposed a socialist transformation of work, economy, society, and culture as axiomatic for a human
and, thus, postandrocentric world."[66] This postandrocentric world, Gilman
believed, would be brought about by a restoration of female choice.

Beginning with *Women and Economics*, the theme of female choice
served as the foundation for Gilman's feminist theorizing and writing
throughout her long career. In *The Man-Made World or, Our Androcentric
Culture* (1911), Gilman referred to the loss of female choice as "the great
fundamental error of the Androcentric Culture" because it impaired both
natural and sexual selection.[67] As a remedy to excessive sex distinctions,
Gilman urged women and men alike to reread their Darwin and learn the
truth about female choice, a natural phenomenon that had been unnatu-
rally supplanted by men. In 1912, Gilman published a fable entitled "Improving on Nature" to highlight the necessity of female choice of sexual

partners. In this tale, a man awakened Mother Nature to alert her that women were trying to be men. Incredulous, Mother Nature asked to hear from women themselves. She was stunned when in walked "a plump, pink little person; hobbled, stilted, and profusely decorated." Mother Nature asked the woman why she was so little, so meek, and so weak. Each time, the woman answered, "[H]e likes us that way." "I never heard such talk!" exclaimed Mother Nature. "What business has he to do the choosing? That is your place, my dear, and has been since you [were] a Cirriped." Gilman's choice of the "Cirriped," more commonly known as barnacles, as her pre-human example demonstrates her familiarity with Darwin's work and also seems strategic. Darwin wrote two early books about barnacles, and, in *The Descent of Man*, Darwin described the male Cirriped as living "like epiphytic plants either on the female or the hermaphrodite form, and are destitute of a mouth and of prehensile limbs. In these cases it is the male which has been modified, and has lost certain important organs, which the female possess."[68] With this example, Gilman established that the male did not always dominate in nature and perhaps that patriarchy itself was unnatural. When man protested women's right to choose mates, Mother Nature showed him examples of females throughout the animal and insect kingdoms to demonstrate the error of his ways. In the end, Mother Nature advised women, "Develop your brains and muscles; earn your own livings; be bought by no man; and choose the kind with which you wish to replenish the earth."[69] Throughout her life, Gilman published scores of articles, books, and novels, but her critique of the man-made world was, in the words of her biographer Judith Allen, "premised on an economically motivated male overthrow of periodic female sex selection."[70]

Gilman attempted to paint a picture of the kind of world that would result from the conscious resumption of female choice in her first utopian novel *Moving the Mountain* (1911), which was followed by the sequels *Herland* (1915) and *With Her in Ourland* (1916). Drawing heavily on the utopian strategies pioneered by Edward Bellamy in *Looking Backward* (1888), Gilman placed the narrator of *Moving the Mountain*, John Robertson, as a stranger in the unfamiliar near future.[71] Robertson went missing in Tibet from 1910 until 1940. By happenstance, his younger sister Ellen, "Nellie," found him in a remote village and brought him home to New York. Upon his return to the United States, Robertson was shocked by the tremendous changes in American society and within his own family that had taken place in just a few decades. For starters, Nellie was a college president, women and men both wore pants, and women and men worked in whatever careers they chose. Robertson repeatedly expressed confusion regarding these new gender roles, wondering aloud where the women went and how, for example, a pretty woman could have become an engineer.[72]

Upon learning that men no longer used tobacco, drank alcohol, or hunted, Robertson, incredulous, asked how such vast changes could possibly have been accomplished in one generation. His sister replied that women simply refused to marry men who smoked, drank, or killed animals for sport (earlier, she had explained how venereal disease was stamped out in similar fashion). Shocked, the narrator exclaimed, "[Y]ou women are trying to make men over to suit yourselves." "Yes. Why not? Didn't you make women to suit yourselves for several thousand years," replied his sister. "You bred and trained us to suit your tastes; you liked us small, you liked us weak, you liked us timid, you liked us ignorant, you liked us pretty—what you called pretty—and you eliminated the kinds you did not like."[73]

While *Moving the Mountain* could be read as a eugenic novel (indeed, several passages discuss the weeding out of undesirable traits through conscious female selection), Gilman justified the changes in mate selection in terms of the benefits derived by existing women, rather than by the benefits that might accrue to future children, and insisted that women themselves were the only ones who should make reproductive decisions. As Nellie, the heroine of *Moving the Mountain*, explained, "the world has come alive. We are doing in a pleasant, practical way, all the things which we could have done, at any time before—only we never thought so. The real change is this: we have changed our minds."[74] The "chief factor" in the change, according to Nellie, was the "individuation" of women: "We individualize the women—develop their personal power, their human characteristics—and they don't have so many children."[75] Among feminist socialists, "individualizing the women" was a common refrain since they rejected arguments based on the rhetoric of race suicide, as well as any suggestion of government or other interference in women's reproductive lives.

For the educated, white women to whom Gilman wrote, her suggested reforms would have produced radical changes, as evidenced by their forceful rejection among mainstream Americans and many of her fellow socialists. While the majority of American socialists did not see eye-to-eye with Gilman or Gamble on the question of feminism or female choice, a vocal minority of transatlantic socialists shared their conviction that women's rights and socialist revolution were twinned causes—to this group of reformers the Darwinian concept of female choice represented the ideal

merger of feminism and socialism. Unlike Gilman who feared, at least initially, that separating sex from reproduction would further unleash male
lust, these reformers believed that the natural, logical extension of female
choice was the movement that came to be called "birth control" under the
leadership of the American socialist Margaret Sanger.

## THE DARWINIAN ORIGINS OF THE
## BIRTH CONTROL MOVEMENT

The socialist and feminist embrace of female choice can especially be
seen in the prehistory of the organized birth control movement in America and in the intellectual development of its founder, Margaret Sanger
(1879–1966). The sixth of eleven surviving children, Sanger grew up seeing
firsthand the perils of too many births. Her mother died at the age of forty-
eight, which Sanger attributed to her having been pregnant eighteen times
during her thirty-year marriage.[76] Sanger dedicated her book *Woman and
the New Race* (1920), which popularized her birth control ideology and became a best seller, to "the memory of my mother, a mother who gave birth
to eleven living children."[77] Sanger's work as a nurse in New York City
at the turn of the twentieth century further convinced her that female
reproductive autonomy through economic independence and scientific advancement was the only surefire path to healthy families and emancipated
women. Growing up, Sanger's father was a freethinker who even invited
Robert Ingersoll to speak in her town. Sanger remembered escorting Ingersoll to a patch in the woods, after they had been barred from the lecture hall in town, and the opprobrium that followed her family for having
brought "the great infidel" to speak.[78] Apparently, the iconoclastic position suited Sanger as she continued to join and lead causes considered radical and infidel for the rest of her life. On New Year's Eve 1913, Sanger had
an epiphany that it was time to dedicate her life to the cause she named
"birth control."[79] After forming the National Birth Control League and determining to publish a magazine called the *Woman Rebel*, Sanger sought
support from other feminists, starting with the most famous feminist in
the world, Charlotte Perkins Gilman, but struck "no responsive chord
from her." Sanger then turned to her friends in the Socialist Party who
sent in hundreds of subscriptions to her new magazine.[80] She had found
her audience and her cause.

American women socialists, much more so than male socialists, were
drawn to reformist readings of evolution and to the possibility of objective
science. According to Mark Pittenger's research on evolutionary thought

in American socialism, male socialists tended to use evolutionary rhetoric
as a way to justify the status quo and, particularly, the outsider status of
women and African Americans.[81] To female socialists, however, evolutionary science offered an appealing alternative to the status quo by redefining
the natural and women's role in it. Between 1901 and 1915, socialist women
spoke and wrote about the feminist implications of evolutionary science—
most often *The Descent of Man* and the work of Lester Frank Ward—in the
periodical *Socialist Woman* and also in the Woman's National Committee
of the Socialist Party. Indeed, the very first issue of the *Socialist Woman*
(June 1907) included a note instructing subscribers to "Read Darwin's 'Descent of Man.' It will give you a pretty good idea of the part the feminine
principle has played in the animal kingdom." And it appears that readers
took this advice because the October 1907 issue included an article about
the Cook County (Illinois) Ladies' Branch's recent decision to embark on
a study of woman in society, which began with a session on "Woman in
Evolution."[82] Female socialists believed that evolutionary theory revealed
the authoritative role that mothers had once played in reproductive decisions and suggested women should regain this power. Articles in the *Socialist Woman* decried economic marriage, criticized church and state for
regulating marriage, and often compared marriage to prostitution. For so-
lutions to the marriage problem, the women writers looked to the animal
kingdom and to Darwinian sexual selection. As Sara Kingsbury noted in
"The Lady-like Woman: Her Place in Nature" (1908), the modern ladylike
woman "violates the habit of every other female in the animal kingdom. . . .
She is the only female in the animal kingdom who seeks to charm the
male." She then observed that Darwin, in *The Descent of Man*, had established the natural pattern of courtship whereby the males must earn
access to the females. "In the animal world there is no economic dependence on the part of the female to drive her to accept the advances of the
amorous male, whether she desires them or not." Kingsbury ended on a
hopeful note, however, observing that "there are those of us who are awakening. We have science for an ally."[83] In addition to the *Socialist Woman*,
readers and contributors to the *Masses*, the avant garde magazine popular
among the Greenwich Village radicals with whom Sanger associated in
the early 1910s, also frequently wrote about female choice, sexual selection, and Lester Frank Ward's gynaecocentrism.[84]

In 1908, however, the tone of *Socialist Woman* shifted from theoretical
to practical affairs, and in 1915, the party disbanded the Woman's National
Committee.[85] Nevertheless, Sanger would have likely read the *Socialist
Woman* and the *Masses* and been familiar with the version of evolution

by female choice promoted by the feminist socialists, including Gilman. Moreover, the readers of *Socialist Woman* were likely many of the same women who subscribed to Sanger's *Woman Rebel*, because by 1914 when her magazine debuted Sanger was well-known in the socialist movement. In 1910, Sanger and her first husband Bill moved from the suburbs to Manhattan in large part to join the socialist and reformist causes flourishing in and around Greenwich Village. In 1911, Margaret joined Branch 5 of the Socialist Party of New York and became a paid organizer for the newly established New York Women's Committee for Propaganda for Socialism and Suffrage. When she was not recruiting female socialists, she campaigned on behalf of her husband who ran for municipal alderman on the Socialist ticket.[86] The Sangers joined the Socialist Party at the peak of its popularity in America, and, as Sanger biographer Jean Baker notes, Sanger remained a committed socialist throughout her life—eventually becoming the only member of her upscale Fishkill, New York, precinct to consistently vote for the Socialist Party candidate for president.[87] In 1912, Eugene V. Debs ran (again) for president as a Socialist and garnered a record six percent of the vote. In addition, that year over one thousand Socialists were elected to office nationwide, including fifty-six mayors and one congressman, but not Bill Sanger.[88] While Margaret Sanger was initially recruited to enlist Socialist women for suffrage, she soon realized that the vote would do little to alleviate the burdens placed on working women unless they could also control their reproductive lives.

When Sanger's husband lost his job as an architect and decided to focus on his painting, she worked as a nurse to help support the family. In her visits with immigrant and poor families on the Lower East Side of Manhattan, Sanger saw women's lives torn apart by frequent pregnancies and babies they could not afford to take care of, and she witnessed many women and children die as a result of these hardships. These experiences crystallized in the story of Sadie Sachs (who may have been a composite of the many women she encountered) that Sanger repeated throughout her career as the reason why she started the birth control movement. Sadie was a young Jewish immigrant whom Sanger nursed through the complications that arose from a self-induced abortion. Sadie told Sanger that she simply could not afford another baby and that she had begged her doctor for reliable contraceptive information, only to be advised that she should tell her husband "to sleep on the roof." Sadie survived her first abortion-related illness, but she was not so fortunate three months later. Sanger was called back to the Sachs residence, only to find Sadie dead following a second self-induced abortion.[89] Sanger turned the story of Sadie and others like her into a political program, and began lecturing about women's sexual health to socialist women's groups.

Sanger's talks on sexual health soon turned to writing. From 1912 to 1913, she wrote a column in the socialist daily paper, the *Call*, entitled "What Every Girl Should Know." Then, in March of 1914 she began publishing the *Woman Rebel*. The paper was shut down in October 1914 for violating obscenity laws prohibiting the mention of birth control or sex education, but the articles that did manage to get published evince a strong faith in both socialism and science. In contrast to the church's and the state's idea of marriage, which *Woman Rebel* columnists equated with prostitution, science offered a positive alternative by suggesting that sex, including even sexual pleasure for women, was natural and should be regulated by women. Furthermore, thanks to promising new technological and medical advancements, science might allow women to separate sex from reproduction all together.[90] Sanger also drew on the well-established arguments for "voluntary motherhood" that had been advanced by female reformers since the 1870s. As Linda Gordon establishes in her authoritative history of birth control in America, suffragists, moral reformers (mainly temperance advocates), and free love advocates disagreed on many things but agreed on voluntary motherhood, which essentially meant the right of women to determine the timing of intercourse with their husbands. To this diverse group of reformers, voluntary motherhood blended traditional ideas about the sanctity of the family and the inherent maternal instinct with a more progressive acceptance of female sexuality but, notably, did not include calls for contraceptives.[91] To the established idea of voluntary motherhood, Gamble, Gilman, and Sanger added the evolutionary concept of female choice, which made such demands seem natural and which, eventually, provided the scientific justification for birth control.

While Sanger knew intuitively and experientially that female reproductive autonomy, evolutionary science, and female economic independence were linked, she did not put the intellectual pieces of the puzzle together until she spent several months in England in 1914–1915, fleeing prosecution in the United States for writing about sex in the *Woman Rebel*. Further testifying to the close links between the American and European movements for birth control, Sanger returned several additional times in the postwar years, meeting frequently with British and European leaders, searching for new birth control technologies, and sharing ideas.[92] In 1914, Sanger's goal in leaving the United States was not only to avoid imprisonment for violating antiobscenity laws but also to learn about the transatlantic history of birth control so that she could craft the most thor-

ough defense possible. In the United States, the Comstock laws had classified all sexual health and birth control materials as obscene since the
1870s, so it was difficult for Sanger to study the history of the movement,
let alone meet its transatlantic leaders.[93] Two intellectual constellations
shaped Sanger's formative experiences abroad in 1914–1915 and her later
ideas about birth control: (1) meeting C. [Charles] V. [Vickery] and Bessie
Drysdale, leaders of the British Neo-Malthusian League; and (2) her vast
reading on the history of birth control, sexology, and reproductive health
at the British Museum under the tutelage of pioneering sexologist Havelock Ellis. Both of these constellations also further tied Sanger's ideas
about birth control to Darwinian evolution.

Upon embarking for England in the fall of 1914, Sanger recalled that her
"keenest desire" was to meet the Drysdales.[94] From the Drysdales, Sanger
learned she was not alone in her belief that reproductive autonomy was the
key to any significant change in women's lives and part of any meaningful
socialist reform. She relished in their common goals and in finding peers
who understood both her unconventional ideas and her uncompromis-
ing tactics. The Neo-Malthusian League was also invested in Darwinian
ideas, as is evident in their name. Birth control reformers on both sides of
the Atlantic had long reiterated the key tenets of Darwinian evolution, and
the two movements shared a common intellectual ancestor in Thomas
Malthus, the man who inspired Darwin's theory of natural selection by
postulating that population growth was inversely related to individual
survival, meaning that there would always be a struggle for existence.
Because they, too, were concerned with unchecked population growth,
nineteenth-century birth control advocates often referred to themselves
as Malthusians and frequently cited both Malthus and Darwin in their
pamphlets.[95] Early condoms were even referred to as "Malthus caps."[96]

The British Neo-Malthusians considered themselves Darwinian and
Malthusian in equal parts, as demonstrated by their rejection of Malthus's
signature recommendation that couples marry late in life and practice
"continence" or abstinence to avoid having too many children. To the
Neo-Malthusians, abstinence and late marriage were impractical suggestions because "experience has shown that the instinct of reproduction is
too strong to admit of this remedy being efficacious."[97] With the *Origin of
Species* and *The Descent of Man*, Darwin successfully argued that man
was an animal, who by extension had natural sexual urges, further proving
to these reformers that abstinence was a futile solution to the problem of
population control. Instead of abstinence, the Neo-Malthusians proposed
early marriage and frequent intercourse with reproductive controls put in

place to avoid having more children than a family wanted or could afford, suggestions that they felt were in line with Darwinian theory. The Neo-Malthusians argued that birth control would make obsolete the brutality of natural selection by ensuring that populations did not grow so large as to force individuals to compete for resources. As C.V. Drysdale explained, "[T]he Malthusian-Darwinian doctrine completely destroys the illusion of any benevolent dispensation [of nature], and reveals Nature as completely indifferent or even hostile to human welfare," which he believed granted credence to their claim that birth control could work as a more benevolent form of natural selection.[98] Ultimately, the Neo-Malthusians hoped any selective pressure would come not from natural selection but from sexual selection—from women, as well as men, intelligently selecting mates and consciously planning the number of wanted children.

The influence of Darwin on the emerging birth control movement was publicly demonstrated during the famous "Fruits of Philosophy" trial in England in 1877, which galvanized the movement on both sides of the Atlantic. In 1876, Charles Bradlaugh and Annie Besant began printing and selling the birth control pamphlet "The Fruits of Philosophy," written years before by the American doctor Charles Knowlton. In 1877, Bradlaugh and Besant were arrested and charged with obscenity, the first trial of its kind in England, which had no equivalent of the U.S. Comstock laws. The jury found the pair guilty of "defaming the morals of the public" and sentenced them to pay a fine and serve jail time, but their conviction was overturned on appeal, establishing the legal precedent that birth control was not, in fact, obscene in England.[99] The massive publicity surrounding the trial also reenergized the British Malthusian League (founded in 1861), which was subsequently known as the Neo-Malthusian League. "In time to come," as C.V. Drysdale later recalled, "the Bradlaugh Besant trial . . . will be recognised as the greatest event in the history of humanity. It heralded the dawn of a new biological era—of the final ascent of Man from the ape." This "final ascent" was predicated on man no longer being subjected to "the terrible internal force of his own blind reproduction." Drysdale further credited the trial with bringing international attention to the Malthusian League, which "led to the translation and distribution of hundreds and thousands of copies of the Knowlton pamphlet and other neo-Malthusian publications all over the world, as well as to the successful launching of . . . [Malthusian] leagues in many other countries and of the modern Birth Control movement."[100]

In pleading their case, Bradlaugh and Besant wanted to establish that birth control was natural and that it accorded with evolutionary principles regarding population growth. To establish the movement's scientific legitimacy, Bradlaugh and Besant hoped especially to call one witness to the stand: Charles Darwin. The Neo-Malthusians believed that overpopulation presented a major problem for society and that it was far better to provide a check on birthrates rather than have death rates escalate. They hoped that Darwin would read excerpts from his books in the courtroom because they felt confident that his theories provided irrefutable evidence for the scientific basis of the birth control movement and that his scientific authority would lend credibility to their cause.[101] Darwin politely asked to be excused from testifying on account of his ill health and the fact that he disagreed with artificial checks to fertilization. In his very cordial reply to Bradlaugh's invitation, Darwin explained that he opposed birth control because he feared "any such practices would in time spread to unmarried women & w$^d$ destroy chastity, on which the family bond depends; & the weakening of this bond would be the greatest of all possible evils to mankind."[102] His response reveals that his objections to birth control were more cultural than scientific. He did not say that birth control failed to accord with evolutionary principles, simply that decoupling sex from reproduction would destroy the bonds of the family, a nuance that the next generation of evolutionists revisited.

Whether he approved or not, Darwin's theories of natural and sexual selection, especially female choice, provided a major intellectual impetus for the birth control movements in England and the United States, as Sanger discovered during her crash course at the British Museum in 1915.[103] Sanger's impending trial in the United States had been compared to the Fruits of Philosophy trial, and she expressed shock that she had to go to all the way to England to learn about it, noting in her autobiography, "I had not found a trace of this in my previous research, even in Boston where [the "Fruits of Philosophy" pamphlet] had been published."[104] While Sanger enjoyed going to art museums and concerts "each week day, however, found me at the British Museum, going in with the opening of the gates in the morning." She had set a tremendous goal for herself, as she explained:

My aim was to present my case from all angles, to make the trial soundly historical so that birth control would be seriously discussed in America. Therefore, I read avidly and voluminously many weighty tomes, and turned carefully the yellowed, brittle pages of pamphlets and broadsides, finding much that was dull, much that was irrelevant, but also much that was amusing, if only for the ponderous manner of its expression. In the end I had a picture of what had gone before.[105]

Fig. 4.2. Margaret Sanger, during her federal trial for obscenity, 1916.
Reproduced by permission from Corbis. ©Bettmann/CORBIS.

Judging from the massive collection of historical materials she amassed, materials that are now part of the "unfilmed collection" of her archive at the Sophia Smith Collection, Sanger learned quite a lot on this trip, and she continued to consult and build her birth control library throughout her long career (fig. 4.2).

For advice on what and whom to read at the British Museum, Sanger consulted Havelock Ellis whose multivolume *Studies in the Psychology of Sex* had impressed her as the best and most scientific account of human sexuality ever written. Sanger first met Ellis just before Christmas 1914, and the two remained close friends, and occasional lovers, throughout their lives. Sanger referred to him as the "King," supported him financially, and was one of only two people to visit him in his last years.[106] Sanger revered Ellis for his frank writings on sexuality, and she described his "great contribution" to the world as being able to "clarify the question of sex, and free it from the smudginess connected with it from the beginning of Christianity, raise it from the dark cellar, set it on a higher plane."[107] Sanger recalled that he "aided me immensely in my studies by guiding my reading. Tuesdays and Fridays were his days at the British Museum, and he often left little messages at my seat, listing helpful articles or offering suggestions as to books which might assist me in the particular aspect I was then engaged upon."[108] Of their relationship, Sanger recalled, "I have never felt about any other person as I do about Havelock Ellis. To know him has been a bounteous privilege; to claim him friend my greatest honor."[109] Years later, her son Grant Sanger emphasized in an interview that since his mother did not have the benefit of college, her "true formal education . . . took place in the year 1914–1915 when she studied in the reading room of the British Museum with Havelock Ellis for her tutor."[110]

Ellis's views about human sexuality directly informed Sanger's and, thus, merit a closer look. According to Sanger biographer Ellen Chesler, "[I]t is virtually impossible to overestimate the impact Ellis would have on Sanger."[111] To Ellis, human sexuality was an important evolutionary activity that should be studied scientifically, as one would study any other aspect of nature, by accumulating data and writing about it objectively, not judgmentally, just as Darwin had done with other aspects of the human experience in *The Descent of Man* and just as Alfred Kinsey would do with human sexuality a generation later. Ellis's ideas about human sexuality were inspired and directly informed by *The Descent of Man*. Indeed, no one did more to popularize the connections between Darwinian evolution and the scientific study of sex than Ellis. In *Studies in the Psychology of Sex* (1896–1928), especially volume four, *Sexual Selection in Man*, Ellis lauded Darwin's contributions to the emerging field of sexology but attempted to pinpoint more precisely the exact mechanisms of sexual selection. Throughout this landmark study, Ellis took Darwin's doctrine of sexual selection as his inspiration and point of departure. Ellis discussed *The Descent of Man* on the very first page of the preface to volume four, noting, "Darwin injured an essentially sound principle [sexual selection] by introducing into it a psychological confusion whereby the physiological sensory stimuli through which sexual selection operates were regarded as equivalent to aesthetic preferences." The problem, according to Ellis, was that Darwin had implied that sexual attraction was an aesthetic phenomenon determined by sight. To Ellis, attraction went far beyond the visual. He contended that "[w]hen we look at these phenomena [those inducing sexual arousal] in their broadest biological aspects, love is only to a limited extent a response to beauty; to a greater extent beauty is simply a name for the complexus of stimuli which most adequately arouses love."

He attempted to decode these stimuli as they appealed to the four senses of touch, smell, hearing, and vision. Studying sexual attraction in this comprehensive way revealed that "sexual selection is no longer a hypothesis concerning the truth of which it is impossible to dispute; it is a self-evident fact."[112] Even though Ellis contended that he was taking issue with Darwin because they disagreed on what exactly elicited sexual attraction (visual preferences or a combination of senses), *Studies in the Psychology of Sex* ultimately reified sexual selection as the most important process in the natural world and made the theory the core principle of sexology.

With Ellis as her tutor, Sanger would also have been instructed to read the work of other naturalists who were engaging with the science of sexuality, largely as a result of the questions raised by *The Descent of Man*. The relationship between evolutionary theory and scientific, as well as popular, understandings of sex is both foundational and generally unspoken. The methodological and ideological links between evolutionary science and sexology, however, are as old as the two fields themselves, tracing back to sexual selection theory.[113] As the British researcher John Allen Godfrey pointed out in his 1901 treatise, *The Science of Sex*, evolution both legitimated and necessitated the scientific study of sex. Godfrey began his book by emphasizing the desirability of a science of sex and love, despite objections from those who preferred to think of them as spiritual and mysterious. He noted that, contrary to the "religious conception of man's nature," scientific research since the publication of the *Origin of Species* indicated that "every emotion, every intellectual faculty of man, has its counterpart in a less developed form in the lower animals."[114] "Therefore it is only by accepting man as an entirely natural organism that we can really construct a science of human sex," observed Godfrey. "Evolution gives us the right to accept him as such, to treat even his loftiest thoughts and feelings as natural phenomena, to bring them within the web of cause and effect." Anticipating later arguments advanced by sexologists, Godfrey contended that "the scientific man is therefore entitled to investigate all the phenomena of sex, and deal with them as he does with all other facets of nature."[115]

It is likely that Sanger's reading list from Ellis also included another influential scientific work on sex inspired by *The Descent of Man*: Patrick Geddes and J. Arthur Thomson's aptly titled *The Evolution of Sex* (1890), published as part of a series on sex that Ellis edited. The first chapter of *The Evolution of Sex* dealt exclusively with Darwin's theory of sexual selection and the opposition to it, taking both its tone and illustrations directly from *The Descent of Man*. Geddes and Thomson noted that Darwin had offered "the most extensive treatment" of secondary sex characteristics to date and, thus, the "reader must be assumed to make reference" to *The Descent of Man* while reading their work. While Geddes and Thomson opposed key elements of sexual selection, including Darwin's claim that it gave rise to secondary sex characteristics, they were nevertheless prompted by his work to "supplement" it with their own theory of sexual development.[116] The crux of their theory was that secondary sex characteristics were not secondary at all, but primary—the result of the essential maleness or femaleness of each individual germ cell. Laden with Victorian notions of proper gendered behavior, their theory of sex differences argued that men, like sperm, were essentially "katabolic" (active, dominated by destructive processes), while women, like eggs, were essentially "anabolic" (passive, dominated by constructive processes). Previous historians of gender and science have rightly focused on the misogynistic elements of this theory and its ramifications, but Geddes and Thomson's work also evidenced a sea change in thinking about human sexuality.[117] In *The Evolution of Sex*, and subsequently in *Sex* (1914), Geddes and Thomson sought to change the tone of discussion about sexuality in order to make it more scientific.

In *Sex*, Geddes and Thomson amplified their thesis that human sexuality was best understood in terms of animal mating and they advocated for a more rational, scientific, and capacious approach to sex education and even birth control. Ignorance, not scientific study, led to sexual deviance and overindulgence, they claimed. The cure, then, was not less talk of sex, but more, especially in biological terms:

> We wish to make our position in this respect quite clear. Through and through, and back to the ovum, Man is a mammal, with a mammal's structure and functions, development and pedigree, with a mammal's strength and weaknesses . . . there is specificity through and through; yet there is a common ground of protoplasm that makes the whole world kin; and Man cannot disown his mammalian ancestry. He is in solidarity with the animal creation and with mammals in particular.[118]

Geddes and Thomson staked future progress and enlightenment on men and women embracing their mammalian ancestry and all its lessons.

Even though Darwin opposed artificial checks to population growth, many among the next generation of evolutionists supported birth control. Geddes and Thomson, for example, grudgingly advocated "the use of artificial preventive checks to fertilization" among married couples, though they preferred "temperance."[119] In keeping with Neo-Malthusian thought, they believed that quality not quantity of offspring was most important from an evolutionary perspective. "The future is not to the most numerous populations, but to the most individuated," declared Geddes and Thomson. To accomplish this goal, Geddes and Thomson demanded a "new ethic of the sexes." Central to this new ethic was the "increasing education and civism of women,—in fact, an economic of the sexes very different from that nowadays so common."[120] Geddes and Thomson dismissed religious concerns about "the idea of the biological control of life," arguing instead that "it is Man's prerogative to use science so that he may enter more and more fully into possession of his kingdom."[121] Clearly, part of Man's kingdom belonged to a new type of woman, and part of this new scientific utopia would be greater female economic and reproductive autonomy. A 1930 history of the birth control movement, which Sanger had in her personal collection, credited the "saner views of sex questions" promoted by Ellis, Geddes, and Thomson for "greatly help[ing] . . . birth control propaganda."[122]

In addition to his efforts to promote the scientific study of sex, Ellis also popularized female choice and linked it to his support for socialism. In an 1887 article, Ellis observed that among humans male choice "clearly had its origin in economic causes," not evolutionary precedent. He further argued that male choice had led to unhappy marriages and to prostitution, but he believed "the economic independence of women can alone place the sexual relationships on a sound and free basis" and "restore to sexual selection its due weight in human development."[123] Later, Ellis blamed the loss of female choice for "the unnatural and repressive influence on the erotic aspect of woman's sexual life" because male choice removed the important elements of courtship and foreplay, privileging the "reproductive side of woman's sexual life" over the pleasurable aspects.[124] Throughout their time together, Ellis validated scientifically Sanger's suspicion that a woman's sexual urges could be as strong and as natural as a man's, which was important to her emerging efforts to decouple sex from reproduction. As her biographer Jean Baker observes, "What she had considered as possibility, he sustained as a scientist."[125] From reading and discussing Ellis's books, Sanger likely concluded that sexual passion was natural, that sexual selection drove the evolutionary process, and that women should play the key role in reproduction (though exactly how was up for debate).

Over the years, Ellis's enthusiasm for female economic independence

seems to have waned. Ellis still wrote in favor of women's rights and socialism, but he agreed with the Ellen Key school of thought, described
in chapter 3, that women must be considered mothers first, humans second.[126] For example, he supported state pensions for mothers, which socialist feminists Sanger and Eliza Burt Gamble rejected. Sanger, too, was certainly influenced by Ellen Key (another Ellis disciple) and her celebration
of female sexuality, but, unlike Ellis and Key, Sanger did not think that
motherhood should be a woman's singular function in life. Rather, in line
with Gilman, with whom she initially quarreled over the topic of birth
control, Sanger believed that women should be considered people first and
allowed to enjoy the full range of human activities, a range that would
be immeasurably broadened if women could control the timing and frequency of pregnancy.[127]

The tension over whether to privilege the individuation of women or
the number of their children, the essence of the Gilman-Key debates analyzed in chapter 3, also created divisions in the early birth control and eugenics movements. At the turn of the twentieth century, individuals and
groups across the political spectrum, including Sanger, advanced eugenic
arguments and invoked the word "eugenics," often in contrasting ways. As
Jesse Battan argues in his study of free love eugenics, "[I]t is a mistake to
associate turn-of-the-century eugenics only with reactionary nationalism,
imperialism, or the celebration of patriarchal monogamy. Racial improvement, as well as the politics of maternalism, appealed to a wide range of
groups on the left as well as the right."[128] At the turn of the twentieth
century, "eugenics" was a messy and wide-ranging movement, encom-
passing individuals and groups that did not agree on much except that the
children of the future should be "better born." Eugenic proposals ran the
gamut from pleas to let women freely select their sexual partners, to requiring health certificates for those seeking to wed, to involuntary ster-
ilization of people deemed "undesirable." Much of the historical research
on eugenics initially focused on state- and institution-based programs, but
more recently scholarly attention has turned to the varied, and often conflicting, ways that eugenic arguments were advanced.[129] In the early years
of the birth control movement, Sanger occasionally used the term "eugenics," joined the nation's most prominent doctors and political leaders in
praising the notorious—and, to modern readers, abhorrent—*Buck v. Bell*
decision of 1927, and attended eugenic conferences when she thought doing so would bolster the cause of birth control, but she was never accepted
by the leading eugenicists (who dismissed birth control as immoral and

who prioritized future babies over the autonomy of women), nor was she an enthusiastic eugenicist herself after her brief foray with the movement in the 1920s.[130]

In place of the organized eugenicists' rhetoric about the fitness and number of offspring, feminist socialists and Neo-Malthusians drew on the Darwinian concept of female choice to scientifically justify their interest in prioritizing the needs of mothers over those of unborn children and their objections to governmental interference in individual reproductive decisions. In the pamphlet "Neo-Malthusians and Eugenics," C.V. Drysdale explained Neo-Malthusians "aimed at eliminating the struggle for existence entirely" by "freeing and educating women as much as possible, in order that they should be able to exercise the fullest sexual selection, and to refuse motherhood whenever they felt impelled in the interests of their children or themselves to do so."[131] To Neo-Malthusians, the key statistic was not birthrate but rate of survival. No factors contributed more to the rate of infant survival than the health of the mother and the extent to which she was economically and emotionally prepared to welcome a baby; hence their argument for female autonomy over rate of reproduction.[132] Sanger took her cue from the British Neo-Malthusians and made women's reproductive autonomy, grounded in economic independence, the cornerstone of her campaign for birth control, arguing that women of all classes and races benefited from planning their pregnancies according to their personal and family needs.[133]

After her time studying in England and Europe, Sanger returned to the United States to open her first birth control clinic—indeed the nation's first birth control clinic—in 1916. In 1920 she published *Woman and the New Race* to explain her movement and what she hoped would result from women being able to select their mates, free from economic necessity, and control their own reproductive lives. Written just as the First World War drew to a close, Sanger felt her message was especially pertinent because, like her Neo-Malthusian colleagues, she believed overpopulation inevitably led to war. As her biographer Jean Baker clarifies, Sanger's use of the term "race," in the title and throughout the book, referred to "society, and humankind," not to a specific racial group.[134] With regard to just what exactly this new race might be, Sanger detailed statistics regarding the ethnic composition of America, immigration, poverty, and disease, not to disparage immigrants and people of color (though some of her dated language might at first seem that way) but to emphasize that "the immigrant brings the possibilities of all these things [art, music, letters, science, courage, and philosophy] to our shores" but that once immigrants arrive they are sequestered in overpopulated slums, given below-subsistence wages, and few opportunities.[135] According to Sanger, the way to keep the birthrate manageable and spread opportunities throughout the population was to teach poor and working-class women, including African Americans, what wealthy women already knew: how to control reproduction.[136] If society enabled motherhood to be free, the "American race, containing the best of all racial elements, could give to the world a vision and a leadership beyond our present imagination."[137]

Throughout *Woman and the New Race*, the Darwinian influences of Sanger's time with the feminist socialists in Greenwich Village and her later experiences in London with Havelock Ellis and the British Neo-Malthusians are evident. In his preface to the book, Ellis began by comparing the woman movement to the labor movement, explaining that while the two had not always worked together in the past, now women "held the secret without which labour is impotent to reach its ends." As the "regulator of the birthrate," women could control, among other things, "the health and longevity of the individual, the economic welfare of the workers, the general level of culture of the community, [and] the possibility of abolishing from the world the desolating scourge of war."[138] In the final chapter of the book, entitled "The Goal," Sanger expounded on the principle of female choice in much the same way that Eliza Burt Gamble and Charlotte Perkins Gilman had a generation earlier: "it is the essential function of voluntary motherhood to choose its own mate, to determine the time of childbearing and to regulate strictly the number of offspring. Natural affection upon her part, instead of selection dictated by social or economic advantage, will give her a better fatherhood for her children." Only when women were economically self-sufficient and able to exercise their own free choice of mates could a woman "give play to her tastes, her talents and her ambitions" and "become a full-rounded human being." Echoing the evolutionary ideas popularized by Lester Frank Ward, Gamble, and Gilman, Sanger advised readers to recall that "in all of the animal species below the human, motherhood has a clearly discernable superiority over fatherhood. It is the first pulse of organic life" and through the mothers come "evolutionary progress." "Why is this true of the lower species yet not true of human beings?" Sanger asked. "The secret is revealed by one significant fact—the female's functions in these animal species are not limited to motherhood alone. . . . Through the development of the individual mother, better and higher types of animals are produced and carried forward."[139] By 1920, direct references to Darwin and *The Descent of Man* had dropped out, but his intellectual influence remained prominent

in Sanger's articulation of the evolutionary importance of female reproductive autonomy and of females' full engagement in the life of the species, commonplace phenomena throughout the animal kingdom.

The eventual success of Sanger's birth control campaign owed much to
the Darwinian revolution and the new thinking about gender and sex that
it inspired. What Darwin had unwittingly done with the publication of
*The Descent of Man* was place evolutionary theory at the center of popular and scientific discussions regarding gender and sex, as well as provide
scientific justification for female reproductive autonomy and the academic
study of sex. Women's growing demands for birth control reflected not only
the logic and appeal of "female choice" but also women's faith in science
and their belief that science could shed light on problems of reproduction.
Again and again, in the pages of the *Woman Rebel* and in *Woman and the
New Race*, Sanger, echoing the women analyzed in chapter 2, called on
science to solve the woman question and favorably contrasted science with
religion.

While Sanger began her birth control advocacy in search of contracep-
tion that women could administer themselves, she eventually hitched her
movement to medical science and lobbied for birth control pills that doctors had to prescribe for women.[140] In some ways, then, the ultimate success of the birth control movement might tell us just as much about the
growing cultural authority of science as it does about the campaign for
women's rights. One might further argue that the particular trajectory of
the American birth control movement owed much to the gendered (and racialized) development of the scientific establishment (men on the inside as
scientists, women on the outside as agitators) and to women's enthusiasm
for science, even though such enthusiasm was often unrequited. Drawing
on science, especially evolutionary theory, allowed feminists, socialists,
and sex reformers to claim that their proposals were natural and to attach themselves to the cultural prestige of science, which connoted modernity, research, and truth in contrast to the tradition, moralizing, and
dogma associated with religion. As Eliza Burt Gamble's writings about sex
first revealed, it was a logical step for women to go from contemplating female choice in animals to demanding reproductive autonomy for women.
Gilman, too, was eventually convinced by the logic of this argument and
threw in her lot with Sanger's birth control movement in the 1930s.[141]

Ultimately, as a study of the early intellectual influences of Margaret
Sanger reveals, the path to birth control in America was paved, at least
partially, by the socialist and feminist articulation of the Darwinian concept of female choice and by women's enthusiasm for evolutionary science.

Darwinian theory overturned Eve's curse; inspired women to trust science as an ally; initiated the scientific study of sex; and gave women a credible, scientific concept to draw on in arguing for reproductive autonomy. These developments helped build the foundation for the organized movement for birth control in America and even help to explain Floyd Dell's (who happened to be Margaret Sanger's Greenwich Village neighbor) curious claim that Darwinian science provided the "basis for a successful feminist movement" in the United States.[142]

# Conclusion

In the final decades of the nineteenth century, biblical and scientific gender paradigms clashed, blended, and in some cases reinforced each other in debates about the "woman question." Characterized by their multivalence and authority, both evolutionary rhetoric and biblical verses were called upon as evidence by supporters and opponents of women's rights. Ultimately, however, evolutionary science displaced religion as the arena in which questions of sex difference could be resolved. Charles Darwin's *The Descent of Man, and Selection in Relation to Sex* (1871) was a defining factor in this transition. According to Darwin, humans were animals, reproduction drove the evolutionary process, and sex differentiation expedited development. As a result, popular and scientific thinking about gender shifted to focus on the origins and biology of sex difference, human's relationship to animals, the science of sexual attraction, and the physiology of maternity, all of which were only hazily understood at that time. Despite gaps in scientific knowledge, feminists and antifeminists alike increasingly incorporated evolutionary arguments as evidence for their claims about women's ideal role in society.

Framing the "woman question" in terms of evolutionary theory allowed opponents of female advancement to call upon nature as an ally and describe female inferiority as a biological inevitability and evolutionary necessity. Such traditional views on the relations of the sexes aligned naturalists with clergymen and helped smooth over their differing explanations for the origins of life. Political, religious, and scientific leaders most often drew on Darwinian evolutionary theory to naturalize the status quo and affirm biological determinism. Indeed, arguments from the natural tend to be made in support of the status quo, perhaps because most people think that the status quo is, inherently, natural. Thus, on first glance, nineteenth-century scientific theories of sex difference focused on establishing women's "natural" inferiority; however, the rocky transition from biblical to scientific explanations of gender difference provided some positive benefits for women. Namely, evolutionary science rendered obsolete the lessons of the Garden of Eden and freed women from having to answer for Eve's sin and abide by her curse. As popular acceptance of evolution, together with other modern developments, eroded faith in biblical literalism, religious justifications for women's second-class status lost their cultural authority. Women continued to find powerful inspiration in the Bible, of course, but organized religion did not often create an environment open to feminist intervention. With its emphasis on testable data and objective experimentation, science did, in theory if not always in practice.

Grounding their arguments in terms of what was natural, however, proved to be tricky business for the Darwinian feminists, as the women in this book demonstrate. For example, drawing on Darwinian rhetoric encouraged women to think in terms of racial hierarchies and to focus on the reproductive aspects of women's lives, which inhibited the movement's overall inclusivity and radicalism. Moreover, embracing evolutionary theory's emphasis on women's bodies also, ultimately, limited the scope of the movement and proved to be a difficult foundation for building consensus because women do not utilize or experience their bodies in the same ways. Rather than mimic male scientists' emphasis on biological determinism, however, the Darwinian feminists suggested that the natural world was full of variety and change, which indicated, as Antoinette Brown Blackwell so powerfully argued, that, one day, men could be seamstresses just as easily as women could be sea captains. Other Darwinian feminists, most notably Helen Hamilton Gardener, rejected the false binary between nature and culture, stressing that the two were in fact deeply enmeshed and mutually constituted, as her essay "Sex in Brain" and later brain dissection testified. The innovative strategies employed by the Darwinian feminists—probing the supposed boundaries not only between humans and animals but also between nature and culture, and challenging the scientific establishment to live up to its impartial mandate—provided a template for women and other outsider groups hoping to challenge biological determinism and anticipated more recent feminist arguments.[1]

Nineteenth-century proponents of women's rights were attracted to evolutionary science because, unlike biblical law, it was easily amendable and open to new ideas. As a result, women working on behalf of women's rights in the 1870s and 1880s eagerly incorporated Darwinian arguments, and, after 1890, many feminist intellectuals working outside of traditional suffrage organizations continued to draw on Darwin and *The Descent of Man* for inspiration. This alternate women's rights discourse, most often promoted by socialists and freethinkers, prioritized the concerns of working mothers, fought to make public the misogyny underlying many mainstream religious (and scientific) beliefs, and suggested that female reproductive autonomy would enable women's overall emancipation in ways that the vote never could. Antoinette Brown Blackwell, Helen Hamilton Gardener, Eliza Burt Gamble, and the other women studied in this book were confident that if evolutionists faithfully applied scientific experiments to the woman question, women would be found to be equivalent, maybe even superior, to men. Compared with a religion whose female role models included a virgin mother and an inquisitive woman responsible for the fall of man, women found evolution to be, at least potentially, a more value-neutral thought system. While these women rejected many of Darwin's specific statements about sex differences, they trusted in an objective scientific method and, more importantly, helped make such a method the norm.

Evolutionary theory also inspired proponents of women's rights by allowing them to imagine a world governed by naturalistic and nonpatriarchal laws. Specifically, Darwinian feminists found affirmation in the variety of gender relations, sexual expressions, and power structures in the animal kingdom. They compared animal relationships and labor patterns with human ones and found humans' wanting. Furthermore, the popularity of evolutionary theory impressed upon women the importance of critical thought and made them more comfortable breaking with tradition. This influence can be seen especially in the connections between Darwinian feminists and the freethought movement at the end of the nineteenth century. Within the freethought movement, feminists like Elizabeth Cady Stanton and Helen Hamilton Gardener found receptive audiences and venues for ideas that were no longer welcomed in women's suffrage organizations.

By refuting the idea that women had been cursed by God to suffer in maternity, Darwin also opened up new discussions about reproduction, motherhood, and female domesticity. Rather than accept pregnancy as a debilitating disease, Darwinian feminists instead looked to the animal kingdom and saw examples of healthy, fit pregnancy. As a result, many women began to demand that pregnancy no longer be defined as a disease and suggested that it might instead be evidence of female superiority. Observing the animal kingdom also shed light on the division of domestic labor. If male spiders and birds helped out with domestic chores, should not men as well? Finally, since mothers played the integral role in the future health of offspring, many women began to insist that mothers have the definitive say in all reproductive decisions. While some, such as Ellen Key, took evolutionary arguments about reproduction to mean that women should devote themselves entirely to bearing and raising children, others, including the feminists and socialists chronicled in chapters 3 and 4, believed that motherhood was just one role that a woman might play in her life. They further believed that restoring female choice among economically self-sufficient women would simultaneously elevate the status of women, although they were mainly concerned with the needs of professional women and not necessarily the women whose labor made it possible for women to have professions. After making the link between animal and human reproduction and simultaneously embracing the idea that science could help emancipate women, it was not a far cry for reformers to turn calls for female choice into demands for birth control.

Read in this light, *From Eve to Evolution* charts the preconditions necessary for reproductive autonomy to be conceptualized and successfully articulated as a demand by women. First, women needed to find a way to challenge the long-standing biblical convictions that they were destined to be subservient to men and suffer in pregnancy; second, they had to trust in science and have reason to enlist it as a force for positive change; third, they had to think critically about motherhood and female domesticity and be able to imagine alternatives to patriarchal heterosexual gender roles; and fourth, science had to engage in the study of sexuality and reproduction. Finally, women's arguments for reproductive autonomy were bolstered by the natural precedent of female choice in the animal kingdom and by the popularity of Darwinian evolutionary theory more broadly.

Darwinian evolution, especially as explained in *The Descent of Man*, prompted Americans to think about sex and gender in terms of nature and animals, a change that powerfully shaped twentieth-century developments including the growth of sexology, the gradual acceptance of birth control, and the secularization of feminist thought. These developments remind us that gender shaped the American reception of evolution and that evolutionary science continues to frame ideas about gender, sex, and race. For historians, looking at gender, religion, and evolutionary theory in concert not only helps us better understand the construction of gender and the development of American feminist thought, it also enriches our understanding of the American reception of Darwin, the ongoing controversies over evolution, and the science of sex difference.

After a promising start in the 1870s and 1880s, however, the historical relationship between evolutionary science and feminist thought has been marked by discord, misunderstanding, and exclusion. Throughout most of the twentieth century, feminism and science tended to proceed along two separate tracks (with the possible exception of the development of the birth control pill, depending on how one views the pill's medical history). Although, as the Darwinian feminists realized, the success of women often goes hand-in-hand with their participation in science and with the overall cultural insistence upon objective, inclusive scientific research. If the nineteenth-century Darwinian feminists chronicled in this book left just one legacy, it might be, to paraphrase Maria Mitchell's 1875 presidential address at the Association for the Advancement of Women, science needs women just as much as women need science.

Thankfully, though still in the minority, women are no longer excluded from the ranks of professional science and feminists no longer express antipathy toward science, but the long history of women's exclusion from science has hurt both women and science. To give just one example, it has impressed upon the public that somehow evolutionary theory, like the literal Genesis, confirms women's subordination by men as natural (a fact evidenced and bolstered by the continued popularity of evolutionary psychology in mainstream culture), when, in fact, as the women chronicled in this book understood, evolutionary theory suggests many possible ways to understand sex difference and sexuality. Those interested in countering the claims of biological determinists would do well to recall their turn-of-the-twentieth-century predecessors who saw in Darwinian evolutionary theory an emphasis on variety and change, especially as revealed by human-animal kinship; a new way to understand the symbiotic relationship between nature and culture; and the potential to revolutionize traditional ideas about gender and sex in order to allow for greater female reproductive autonomy, the equitable distribution of domestic and professional labor, and an increased appreciation of all the diversity and wonder found in the animal and plant kingdoms, including among humans.

# ACKNOWLEDGMENTS

Over the many years it has taken me to research and write this book, I have been aided and enlightened by the wise advice, generous feedback, and timely assistance of many scholars, colleagues, friends, librarians, and archival staff at various institutions around the world. It is with deep gratitude that I begin to acknowledge them here. At the University of Texas at Austin, I thrived under the caring mentorship of Bob Abzug, whose guidance, insights, and encouragement continue to influence my professional development. I also want to thank all the members of the University of Texas American Studies Department for providing such an intellectually rich and supportive environment, especially Janet Davis, a true role model and friend. I am grateful as well for the scholarly community I found in the Symposium on Gender, History, and Sexuality, then led by Carolyn Eastman, who continues to be a source of friendship and inspiration. My experiences at the University of Texas were also deeply enriched by the friendship of Elisabeth Horany Carrell, Matt Hedstrom, Sarah Mullen, Nancy LaGreca, Lissa Bollettino, and Erica Whittington.

At its earliest stages, this project was made possible thanks to vital research support from the New England Regional Fellowship Consortium (N.E.R.F.C.); the Sophia Smith Collection at Smith College in Northampton, Massachusetts; and the Sallie Bingham Center for Women's History and Culture at Duke University. The N.E.R.F.C. grant supported research at the Schlesinger Library on the History of Women in America, Radcliffe Institute, Harvard University; the Center for the History of Medicine at the Francis A. Countway Library of Medicine, Harvard Medical School; and the Boston Athenaeum. The Margaret Storrs Grierson Scholar-in-Residence Fellowship allowed me to spend six weeks at the Sophia Smith Collection and Smith College Archives, and the Mary Lily Research Grant

funded two valuable weeks at Duke's Sallie Bingham Center. I would like
to thank the librarians, archivists, and staff at each of these institutions
for so generously sharing their time and expertise with me. In particular,
Amy Hague at the Sophia Smith Collection and Nanci Young of the Smith
College Archives have been incredibly helpful every step of the way, from
suggesting materials in the early months of this project to, more recently,
helping me secure permissions and select the best photos for the book.
While I was at the University of Texas, additional research funding also
came from the Austin branch of the American Association of University
Women and the P.E.O. (Providing Educational Opportunities) International, as well as from several university grants and fellowships.

Two faculty summer research awards from Miami University of Ohio
enabled me to conduct additional research at the Library of Congress in
Washington, D.C., and at the Kinsey Institute for Sex, Gender, and Reproduction at Indiana University (the Kinsey research ultimately appeared as
a separate article). In addition, I am most grateful for two blissful months
at the Huntington Library in San Marino, California, thanks to the Ken-
neth and Dorothy Hill and Mayers fellowships and to a pretenure research
leave from Miami University. I would also like to thank Barbara Finlay,
who oversees the Burt Wilder Brain Collection at Cornell University, for
allowing me access to the collection; Sheila Ann Dean for helping me lo-
cate and photograph sources at Cornell; the staff at the New York Public
Library for permission to quote from the Carrie Chapman Catt papers; and
the editors of the Darwin Correspondence Project at Cambridge University, especially Rosemary Clarkson, for retranscribing unpublished letters
from Darwin and allowing me to quote from them.

Over the years, I have presented portions of this work at dozens of sem-
inars and conferences where I have learned a great deal from the research,
questions, and comments of my fellow presenters, panel commentators,
and audience members. In particular, I would like to thank Robert Richards and the History of Human Sciences Workshop at the University of
Chicago for inviting me to present my research in 2010; the Max Planck
Institute for the History of Science (Berlin) for inviting me to participate
in the summer 2010 Gender Studies of Science Workshop on women and
science "beyond the academy"; and the Newberry Library's Seminar on
Women and Gender, coordinated by Joan Johnson and Francesca Morgan,
for including me in their 2009 schedule. In 2008, I organized a panel at
the Berkshire Conference of Women Historians featuring Cynthia Eagle
Russett called "*Sexual Science* Revisited" to celebrate her groundbreak-
ing book (1989) and highlight the new work on women and nineteenth-

century science inspired by it. Her participation and encouragement provided crucial intellectual support of this project; thanks also to panelists Carla Bittel and Erika Milam for participating in what turned out to be an exceptionally fruitful session and for their continued collaboration. I also want to thank the members of the American Studies Association's Science and Technology Caucus for encouraging the interdisciplinary study of science, especially Carolyn de la Peña, who cofounded the caucus with me in 2007. I have found an exceptionally supportive network of scholars through the History of Science Society's Women's Caucus and want to thank these women, including Sally Gregory Kohlstedt, Karen Rader, Erika Milam, Georgina Montgomery, and Sarah Richardson, for improving my work in countless ways with their own scholarship, feedback, and questions. Special thanks to Adam Shapiro for introducing me to so many historians of science and encouraging my interdisciplinary approach to Darwin in America.

Other scholars and friends have read all or part of this manuscript and offered invaluable feedback and much needed encouragement. In particular, I owe a huge debt of gratitude to the outside readers of this book who identified themselves to me, Constance Areson Clark, Sally Gregory Kohlstedt, and Christina Cogdell, and to the outside readers who chose to remain anonymous. I greatly appreciate their thoughtful, careful readings of my manuscript and expert suggestions for improving it. Carla Bittel generously read the entire manuscript at its next-to-last stage and offered sage advice throughout the entire publication process. Others read all or part of the book along the way and improved it tremendously with their insights, including Kathi Kern, Wendy Kline, Erika Milam, and Natalie Dykstra. Many of my colleagues in American Studies and History at Miami University have also read all or part of the book and offered helpful feedback, including: Peggy Shaffer, Sheila Croucher, Helen Sheumaker, Kelly Quinn, Oana Godeanu-Kenworthy, José Amador, Mary Frederickson, Judith P. Zinsser, Mary Cayton, Allan Winkler, Drew Cayton, and the members of the American Cultures faculty seminar. Thanks to the generous support of Miami University's American Studies Program, Louise Michele Newman came to campus to lead a manuscript revision workshop for me in 2009, which helped me hone the focus of this book and provided a welcome opportunity to share early drafts with one of the leading experts in the field. I also want to thank Carolyn de la Peña, Siva Vaidhyanathan, and Clark Dougan for their early enthusiasm for this project, extremely helpful feedback on initial drafts, and strong support of the interdisciplinary study of science, technology, and culture.

It has been a joy and an honor to work with the editors and staff at the University of Chicago Press, especially Robert Devens and Karen Merikangas Darling, who have been extremely supportive of my project from day one and who have generously made themselves available to answer my many first-time-author questions. Special thanks also to Abby Collier for her patient and expert assistance during the final stages of production.

No one could ask for more supportive colleagues or more engaged students than those I have found at Miami University. Heartfelt thanks to my faculty mentors and colleagues Peggy Shaffer, Sheila Croucher, Allan Winkler, and Tim Melley, all of whom model what it means to be a teacher-scholar and who demonstrate, every day, the value of the humanities not only in our universities, but, more importantly, in our world. Since joining the faculty at Miami University, I have also benefited from the sharp intellect and beautiful prose style of my writing partner, Cindy Klestinec. Three student research assistants have helped me track down and organize sources over the years: Kate Ely, Michelle Prior, and Stephanie Niro. Warm thanks to them for their excitement about working with primary sources and their enthusiasm for historical research. Thanks also to the staff of the Interdisciplinary Programs Office, especially Cathie Isaacs, for generously helping me at many stages of this project and in various capacities. Besides tremendous faculty and students, Miami University is also very lucky to have such great librarians; special thanks to Jenny Presnell and the rest of the library staff for countless favors and expert answers to many questions over the years.

Finally, I would like to thank my parents, Ray and Kay Hamlin, and my husband, Michael Christner. My parents have enthusiastically supported me in everything I have ever done, and they sparked my interest in history by planning all our family vacations around a stop at a Civil War battlefield or other historic site. My husband Michael had the unfortunate luck of meeting me just a few weeks before I wrote the first proposal for this project, so he has been with me quite literally every step of the way, including spending rainy weekends in archives, countless hours fixing crashed hard drives and uncooperative printers, and even a few weekends at academic conferences. Thankfully, he stuck it out with this book and with me. We now share our home and our lives with our daughter Ruby and, soon, a new baby who will arrive just a few months before this book.

INTRODUCTION

1. Floyd Dell, *Women as World Builders: Studies in Modern Feminism* (Chicago: Forbes and Company, 1913), 44. In the next sentence, Dell mentions social evolutionist Herbert Spencer by name, but this quote is from Dell's chapter on Olive Schreiner and Isadora Duncan, who made explicit references to Charles Darwin later in the chapter, demonstrating, among other things, the American tendency to conflate Darwinian and Spencerian ideas about evolution.

2. Fiona Erskine, "The *Origin of Species* and the Science of Female Inferiority," in *Charles Darwin's The Origin of Species: New Interdisciplinary Essays*, ed. David Amigoni and Jeff Wallace, Texts in Culture Series (New York: Manchester University Press, 1995), 100. For other work in this vein, see Flavia Alaya, "Victorian Science and the 'Genius' of Woman," *Journal of the History of Ideas* 38, 2 (April–June 1977): 261–80; Susan Sleeth Mosedale, "Science Corrupted: Victorian Biologists Consider 'The Woman Question,'" *Journal of the History of Biology* 11 (Spring 1978): 1–55; Janice Law Trecker, "Sex, Science and Education," *American Quarterly* 26 (October 1974): 352–66; Ruth Hubbard, Mary Sue Henifin, and Barbara Fried, ed., *Women Look at Biology Looking at Women: A Collection of Feminist Critiques* (Boston: G. K. Hall, 1979); Ruth Hubbard, Mary Sue Henifin, and Barbara Fried, ed., *Biological Woman—the Convenient Myth: A Collection of Feminist Essays and a Comprehensive Bibliography* (Cambridge, MA: Schenkman Publishing Company, 1982); Marian Lowe and Ruth Hubbard, ed., *Woman's Nature: Rationalizations of Inequality*, The Athene Series (New York: Pergamon Press, 1983); Sue Rosser and Charlotte Hogsett, "Darwin and Sexism: Victorian Causes, Contemporary Effects," in *Feminist Visions: Toward a Transformation of the Liberal Arts Curriculum*, ed. Diane Fowlkes and Charlotte McClure (Tuscaloosa: University of Alabama Press, 1984): 42–52. In particular, see Hubbard's essay "Have Only Men Evolved?" in both *Women Look at Biology* and *Biological Woman*. For a collection of many of the primary source scientific articles in question, see Louise Michele Newman, ed. *Men's Ideas, Women's Realities: Popular Science, 1870–1915*, The Athene Series (New York: Pergamon Press, 1985). The best and most thorough account of the sexist uses of

science in the nineteenth century remains Cynthia Eagle Russett, *Sexual Science: The Victorian Construction of Womanhood* (Cambridge: Harvard University Press, 1989). This book is also indebted to Rosalind Rosenberg's *Beyond Separate Spheres: Intellectual Roots of Modern Feminism* (New Haven: Yale University Press, 1982). See also Rosalind Rosenberg, *Changing the Subject: How the Women of Columbia Shaped the Way We Think about Sex and Politics* (New York: Columbia University Press, 2004); Rosalind Rosenberg, "In Search of Woman's Nature, 1850–1920," *Feminist Studies* 3 (Fall 1975): 141–54; and Rosalind Rosenberg, "The Dissent from Darwin, 1890–1930: The New View of Woman among American Social Scientists" (Ph.D. diss., Stanford University, 1974). For additional studies on science and women in the nineteenth century, see Marie Tedesco, "Science and Feminism: Conceptions of Female Intelligence and Their Effect on American Feminism, 1859–1920" (Ph.D. diss., Georgia State University, 1978); Carroll Smith-Rosenberg and Charles Rosenberg, "The Female Animal: Medical and Biological Views of Woman and Her Role in Nineteenth-Century America," *Journal of American History* 60 (September 1973): 332–56; Jill Conway, "Stereotypes of Femininity in a Theory of Sexual Evolution," in *Suffer and Be Still: Women in the Victorian Age*, ed. Martha Vicinus (Bloomington: Indiana University Press, 1972): 140–54.

3. Beryl Satter, *Each Mind a Kingdom: American Women, Sexual Purity, and the New Thought Movement, 1875–1920* (Los Angeles: University of California Press, 1999), 36–7.

4. Several scholars have written articles or chapters on individual women or incidents chronicled in this book, and I have found their work exceedingly helpful in contextualizing my own, especially Sally Gregory Kohlstedt and Mark R. Jorgensen, "'The Irrepressible Woman Question': Women's Responses to Evolutionary Ideology," in *Disseminating Darwinism: The Role of Place, Race, Religion, and Gender*, ed. Ronald L. Numbers and John Stenhouse (New York: Cambridge University Press, 1999): 267–93. See also, Evelleen Richards, "Darwin and the Descent of Woman," in *The Wider Domain of Evolutionary Thought*, ed. David Oldroyd and Ian Langham (Boston: D. Reidel Publishing Company, 1983): 57–111; Penelope Deutscher, "The Descent of Man and the Evolution of Woman: Antoinette Blackwell, Charlotte Perkins Gilman and Eliza Gamble," *Hypatia* 19 (Spring 2004): 35–55; Rosemary Jann, "Revising the Descent of Woman: Eliza Burt Gamble," in *Natural Eloquence: Women Reinscribe Science*, ed. Barbara T. Gates and Ann B. Shteir, Science and Literature Series, ed. George Levine (Madison: University of Wisconsin Press, 1997): 147–63. For British women's responses to Darwin, see Evelleen Richards, "Redrawing the Boundaries: Darwinian Science and Victorian Women Intellectuals," in *Victorian Science in Context*, ed. Bernard Lightman (Chicago: University of Chicago Press, 1997): 119–42; and Barbara T. Gates, "Revisioning Darwin with Sympathy: Arabella Buckley," in *Natural Eloquence*, 164–76. For a discussion of Darwin and the women's movement in England, see Elizabeth Fee, *Science and the "Woman Question," 1860–1920: A Study of English Scientific Periodicals* (Ph.D. diss., Princeton University, 1978); and Elizabeth Fee, "Science and the Woman Problem: Historical Perspectives," in *Sex Differences: Social and Biological Perspectives*, ed. and with an introduction by Michael S. Teitelbaum (Garden City, NY: Anchor Books, 1976): 175–223. For several related articles on the gendered visual depictions of evolutionary science in British and European contexts, see the essays collected in

Ann B. Shteir and Bernard Lightman, eds., *Figuring It Out: Science, Gender, and Visual Culture* (Lebanon, NH: Dartmouth College Press, 2006).

5. The Darwin and Gender Project, part of the larger Darwin Correspondence Project at Cambridge University, has revealed that Darwin's ideas about gender and, in particular, women's aptitude for education and engagement in science were more nuanced (and favorable) in his personal correspondence than in his published writing. See, for example, the "Top Ten Gender Letters" identified here, http://www.darwinproject .ac.uk/top-10-gender-letters.

6. This book relies on the second edition of Darwin's *The Descent of Man*. Charles Darwin, *The Descent of Man, and Selection in Relation to Sex*, 2d ed. (1879), with an introduction by Adrian Desmond and James Moore (New York: Penguin Classics, 2004), 256. Page citations are to this edition. The first edition came out in two separate books; the second is a singular book. Desmond and Moore argue convincingly that the second edition contains the most developed version of Darwin's theory of sexual selection. For the first edition, see Charles Darwin, *The Descent of Man, and Selection in Relation to Sex* (1871), introduction by John Tyler Bonner and Robert M. May (Princeton: Princeton University Press, 1981).

7. Darwin, *Descent of Man*, 629.

8. As literary scholar Bert Bender argues, the *Descent of Man* "powerfully disrupted the Victorian sense of order by initiating the scientific analysis of sex itself, demystifying it and paving the way for the next generation of modernist sexual theory that began almost immediately in the work of Freud and Havelock Ellis." Bert Bender, *The Descent of Love: Darwin and the Theory of Sexual Selection in American Fiction, 1871–1926* (Philadelphia: University of Pennsylvania Press, 1996), 16. Or, as Lawrence Birken points out, "the Darwinian vision proved to be disturbingly subversive" of the Enlightenment sexual order defined by binary gender roles and "separate spheres" for men and women. Lawrence Birken, "Darwin and Gender," *Social Concept* 4 (December 1987), 77. This argument is further developed in Lawrence Birken, *Consuming Desire: Sexual Science and the Emergence of a Culture of Abundance, 1871–1914* (Ithaca, NY: Cornell University Press, 1988). Frank Sulloway makes a similar argument in *Freud, Biologist of the Mind: Beyond the Psychoanalytic Legend* (New York: Basic Books, 1979). Sulloway suggests that the field of sexology emerged with the publication of the *Descent* and that Freud was more influenced by Darwin than any other intellectual source (238). For another study of the influence of Darwin on Freud, see Lucille B. Ritvo, *Darwin's Influence on Freud: A Tale of Two Sciences* (New Haven: Yale University Press, 1990).

9. The other volumes in the *Beagle*'s library consisted of the Bible, history, travelogues, and natural history. The Darwin Correspondence Project at Cambridge University has posted a list of known books on the *Beagle* that can be accessed at http://www .darwinproject.ac.uk/books-on-the-beagle.

10. R.D. Keynes, ed. *Charles Darwin's Beagle Diary* (Cambridge: Cambridge University Press, 1988), 111. Quoted in Janet Browne, *Charles Darwin: Voyaging, A Biography*, vol. 1 (Princeton: Princeton University Press, 1995), 233, fn. 41. For additional biographies of Darwin, see vol. 2 of Janet Browne's biography, *The Power of Place* (1995); Janet Browne, *Darwin's Origin of Species: A Biography* (New York: Atlantic Monthly Press, 2006); as well as the biographical work on Darwin by Adrian Desmond and James

Moore, including *Darwin* (London: Penguin, 1991) and *Darwin: The Life of a Tormented Evolutionist* (New York: W.W. Norton, 1994). For a more personal take on the development of Darwin's thinking, see Randal Keynes (Darwin's great-great-grandson), *Darwin, His Daughter, and Human Evolution* (New York: Riverhead Books, 2001).

11. See, for example, Francis Buzzacott and Mary Isabel Wymore, *Bi-Sexual Man, or Evolution of the Sexes, Scientific Edition* (Chicago: M.A. Donahue and Company, 1912). The authors argued that since the two sexes had evolved from a single bisexual organism, as established by Darwin, bisexual people were superior because they could take part in all life processes. In addition, Edward Carpenter and other early sexologists drew on evolutionary arguments to reframe discussions of homosexuality in naturalistic, nonjudgmental terms. See, for example, Carpenter, *The Intermediate Sex: A Study of Some Types of Transitional Men and Women* (London: Swan Sonnenschein and Co., Inc, 1908). In 2004, evolutionary biologist Joan Roughgarden published a critique of standard interpretations of sexual selection theory which focus on reproduction, emphasizing instead the variety of nonreproductive sexual relationships and gender identities found in the animal kingdom: *Evolution's Rainbow: Diversity, Gender, and Sexuality in Nature and People* (Los Angeles: University of California Press, 2004). See also, Joan Roughgarden, *The Genial Gene: Deconstructing Darwinian Selfishness* (Los Angeles: University of California Press, 2009) and Timothy Morton's essay "Guest Column: Queer Ecology," *PMLA* 125 (March 2010): 273–82.

12. Londa Schiebinger, *Nature's Body: Gender in the Making of Modern Science* (New Brunswick, NJ: Rutgers University Press, 2004). Thomas Laqueur advances similar arguments in his work; see especially *Making Sex: Body and Gender from the Greeks to Freud* (Cambridge: Harvard University Press, 1990).

13. Darwin, *Descent of Man*, 674–75.

14. Darwin's next book further explored the question of human-animal kinship, Darwin, *The Expression of the Emotions in Man and Animals* (London: John Murray, 1872). In Darwin's final book, *The Formation of Vegetable Mould, through the Action of Worms* (London: John Murray, 1881), he recounted a series of experiments demonstrating that even earthworms display a semblance of rational thought as they learn lessons about how to maneuver different sized twigs into their holes.

15. George Levine, *Darwin Loves You: Natural Selection and the Re-enchantment of the World* (Princeton: Princeton University Press, 2006), 197.

16. For work that analyzes the gendered development of modern science, see, to name just a few examples, Londa Schiebinger, *Nature's Body*; Londa Schiebinger, *The Mind Has No Sex? Women in the Origins of Modern Science* (Cambridge: Harvard University Press, 1989); Carolyn Merchant, *The Death of Nature: Women, Ecology, and the Scientific Revolution* (San Francisco: Harper and Row, 1980); and Judith P. Zinsser, ed. *Men, Women, and the Birthing of Modern Science* (DeKalb: Northern Illinois University Press, 2005).

17. Jackson Lears, *Rebirth of a Nation: The Making of Modern America, 1877–1920* (New York: Harper Collins, 2009) 204, 237. Rebecca Edwards concurs: "[F]or many, evolution was synonymous with progress." Rebecca Edwards, *New Spirits: Americans in the Gilded Age, 1865–1905* (New York: Oxford University Press, 2006), 153. Historian Daniel E. Bender traces how evolutionary discourse, specifically theories of social

evolution, was invoked in defense of industrial progress at the turn of the twentieth century in *American Abyss: Savagery and Civilization in an Age of Industry* (Ithaca, NY: Cornell University Press, 2009).

18. Peter J. Bowler, *The Eclipse of Darwinism: Anti-Darwinian Evolution Theories in the Decades around 1900* (Baltimore: Johns Hopkins University Press, 1983).

19. It would be impossible to list all the books on Darwin in America here. Among the classic studies of Darwin in the United States are Jon H. Roberts, *Darwinism and the Divine in America: Protestant Intellectuals and Organic Evolution, 1850–1900* (Madison: University of Wisconsin Press, 1988); Ronald L. Numbers and John Stenhouse, ed. *Disseminating Darwinism: The Role of Place, Race, Religion, and Gender* (New York: Cambridge University Press, 1999); Ronald L. Numbers, *Darwinism Comes to America* (Cambridge: Harvard University Press, 1998); Cynthia Eagle Russett, *Darwin in America: The Intellectual Response, 1865–1912* (San Francisco: W.H. Freeman and Company, 1976); Edward J. Larson, *Evolution: The Remarkable History of a Scientific Theory* (New York: Modern Library, 2004); Edward J. Larson, *Trial and Error: The American Controversy over Creation and Evolution* (New York: Oxford University Press, 1985); Edward J. Larson, *Summer for the Gods: The Scopes Trial and America's Continuing Debate over Science and Religion* (Cambridge: Harvard University Press, 1997).

20. Donna Haraway's work on primates helped me think about the Darwinian cosmology, especially in terms of its presentation of an alternative creation story. Donna Haraway, *Primate Visions: Gender, Race, and Nature in the World of Modern Science* (New York: Routledge, 1989).

21. Darwin, *Descent of Man*, 43.

22. In addition to the Penguin Classics reprint of the second edition of *The Descent of Man* edited and with an introduction by James Moore and Adrian Desmond (2004), two particularly important books on *The Descent of Man* and sexual selection have been published in recent years: Adrian Desmond and James Moore's *Darwin's Sacred Cause: How a Hatred of Slavery Shaped Darwin's Views on Human Evolution* (New York: Houghton, Mifflin, Harcourt, 2009); and Erika Lorraine Milam, *Looking for a Few Good Males: Female Choice in Evolutionary Biology* (Baltimore: Johns Hopkins University Press, 2010).

23. Charles Darwin, *On the Origin of Species, a Facsimile of the First Edition* (1859) with an introduction by Ernst Mayr (Cambridge: Harvard University Press, 1964), 488.

24. For studies of the impetus and development *The Descent of Man*, see Desmond and Moore, *Darwin's Sacred Cause* (2009); and Robert J. Richards, *Darwin and the Emergence of Evolutionary Theories of Mind and Behavior* (Chicago: University of Chicago Press, 1987), esp. ch. 5.

25. Darwin mentioned sexual selection in both his 1842 and 1844 sketches of evolutionary theory according to Michael Ghiselin, *The Triumph of the Darwinian Method* (Berkeley: University of California Press, 1969), 220. Darwin describes sexual selection in the *Origin of Species*, 88–9.

26. Darwin, *Descent of Man*, 262–3.

27. This quote comes from introductory comments that Darwin included on a

paper by W.T. Van Dyck regarding sexual selection in dogs. This paper was read at the Zoological Society of London on April 18, 1882, the day before Darwin died. Charles Darwin, Preliminary Notice, in W. [William] T. [Thomson] Van Dyck, "On the Modification of a Race of Syrian Street-Dogs by Means of Sexual Selection," *Proceedings of the Zoological Society of London* no. 25: 367–9. Available online via John van Wyhe, ed., The Complete Work of Charles Darwin Online. (http://darwin-online.org.uk/). George Romanes reprinted this quote in "The Darwinian Theory," in *Darwin and after Darwin*, vol. 1 (Chicago: Open Court, 1892), 400.

28. "New Publications: The Descent of Man," *New York Times*, June 1, 1871, p. 2.

29. Francis Darwin, ed., *The Life and Letters of Charles Darwin, Including an Autobiographical Chapter, volume III* (London: J. Murray, 1887), 133; quoted in Gertrude Himmelfarb, *Darwin and the Darwinian Revolution* (Garden City: Doubleday, 1962), 355, fn. 11.

30. Hooker to Darwin, March 26, 1871. Joseph Dalton Hooker, *Life and Letters*, vol. 2, ed. Leonard Huxley (London: J. Murray, 1918), 125; quoted in Himmelfarb, *Darwin and the Darwinian Revolution*, 355, fn. 10.

31. "Literary Notes," *Appletons' Journal: A Magazine of General Literature* 5, May 20, 1871, p. 596.

32. Sidney Ratner, "Evolution and the Rise of the Scientific Spirit in America," *Philosophy of Science* 3 (1936): 113; quoted in Bert Bender, *The Descent of Love*, 3, n. 3.

33. Literary Notices, *Godey's Lady's Book and Magazine* 82 (May 1871): 479.

34. "The Descent of Man," *Galaxy* 9 (March 1871): 463. The *Galaxy* printed "Sea" instead of "Sex" in the full title of the work. Citing this misprint as one example, literary scholar Bert Bender, who examined responses to sexual selection in fiction, noted that most reviews he found were loath to print the term "sexual selection," but I did not find that to be the case generally speaking.

35. "Darwin's Descent of Man," *Old and New* 3 (May 1871): 598.

36. "Darwin on the Descent of Man," *Harper's New Monthly Magazine* (July 1871): 305; "Table-Talk," *Appletons' Journal: A Magazine of General Literature* 5, February 11, 1871, pp. 174–5.

37. See, for example, Augusta Leypoldt and George Iles, ed., *Lists of Books for Girls and Women and Their Clubs* (Boston: American Library Association, The Library Bureau, 1895), 108–11. Copy residing at the Boston Athenaeum. "The Aguilar Free Library," *New York Times*, July 5, 1896, p. 24.

38. "The Darwinian Eden," *Overland Monthly and Out West Magazine* 7 (July 1871): 164. The reviewer also declared the book the "best natural history of modern times."

39. "The Museum," *Appletons' Journal: A Magazine of General Literature* 5, April 15, 1871, pp. 447–8; and "The Museum," *Appletons' Journal: A Magazine of General Literature* 5, April 22, 1871, pp. 479–80.

40. "A Logical Refutation of Mr. Darwin's Theory," *Harper's Bazaar*, May 6, 1871, p. 288.

41. "The Descent of Man," *Harper's Bazaar*, June 28, 1873, p. 416.

42. For a study of the influence of sexual selection on American literature, including but not limited to Kate Chopin, see Bert Bender, *The Descent of Love*. For

Darwinian ideas about gender and sex in British literature, see, for example, Gillian Beer *Darwin's Plots: Evolutionary Narrative in Darwin, George Eliot and Nineteenth-Century Fiction*, 3d ed. (New York: Cambridge University Press, 2009); Jennifer Elisabeth Gerstel, "Sexual Selection and Mate Choice in Darwin, Eliot, Gaskell, and Hardy" (Ph.D. diss., University of Toronto, 2002); Angelique Richardson, *Love and Eugenics in the Late Nineteenth Century: Rational Reproduction and the New Woman* (New York: Oxford University Press, 2003); and Patricia Murphy, "Re-evaluating Female 'Inferiority': Sarah Grand versus Charles Darwin," *Victorian Literature and Culture* 26, no. 2 (1998): 221–36. For American short stories that referenced the *Descent*, see William Black, "Madcap Violet," *Galaxy* 21 (May 1876): 602–8; Margaret Vandegrift, "Mademoiselle Stylites," *Lippincott's Magazine of Popular Literature and Science* 11 (April 1873): 459–64; and Edith Wharton, "The Descent of Man," in *The Descent of Man and Other Stories* (1904; reprint, New York: Books for Libraries Press, 1970): 1–34.

43. Harriet Beecher Stowe, *My Wife and I; or Harry Henderson's History* (New York: J.B. Ford and Company, 1871), 321–2.

44. "The New Woman Speculating on the Descent of Man," *Philadelphia Inquirer*, March 5, 1899, p. 3.

45. *Blackwood's Edinburgh Magazine* ran this spoof of *The Descent of Man* (Darwinian Loquitur) in its April 1871 edition, and it was widely reprinted in the United States. See, "The Descent of Man (Darwinian Loquitur)" in *Appletons' Journal: A Magazine of General Literature* , May 13, 1871, pp. 558–9; "The Descent of Man (Darwinian Loquitur)," *Christian Advocate*, June 22, 1871, p. 194; "The Descent of Man (Darwinian Loquitur)," *The Eclectic Magazine of Foreign Literature* 13 (June 1871): 696–8; "The Descent of Man (Darwinian Loquitur)," *Medical and Surgical Reporter*, October 14, 1871, p. 351; and "The Descent of Man (Darwinian Loquitur)," *The Scientific American*, June 3, 1871, p. 361.

46. *The Fall of Man: Or, the Loves of the Gorillas. A Popular Scientific Lecture upon the Darwinian Theory of Development by Sexual Selection. By a Learned Gorilla* (New York: G.W. Carleton & Co, 1871). Published anonymously by Richard Grant White.

47. White, *The Fall of Man*, 8, 9.

48. This waxing process is described in White, *Fall of Man*, 33–4, and its implications for humans on pages 37–8. For an analysis of the gendered and racialized significance of female facial and body hair in an evolutionary context, see Kimberly A. Hamlin, "The 'Case of a Bearded Woman': Hypertrichosis and the Construction of Gender in the Age of Darwin," *American Quarterly* 63 (December 2011): 955–81.

49. For more on the professionalization and masculinization of science in America, see Margaret W. Rossiter, *Women Scientists in America: Struggles and Strategies to 1940* (Baltimore: Johns Hopkins University Press, 1982), esp. ch. 4. For classic histories of the women's rights and suffrage movements, see Ellen Carol DuBois, *Feminism and Suffrage: The Emergence of an Independent Women's Movement in America, 1848–1869* (Ithaca, NY: Cornell University Press, 1978); Ellen Carol DuBois, *Woman Suffrage and Women's Rights* (New York: New York University Press, 1998); Eleanor Flexnor, *Century of Struggle: The Woman's Rights Movement in the United States* (Cambridge: Harvard University Press, 1996); Nancy F. Cott, *The Grounding of Modern Feminism* (New

Haven: Yale University Press, 1987); and Aileen S. Kraditor, *The Ideas of the Woman Suffrage Movement, 1890–1920*, reprint of 1965 edition (New York: Norton, 1981). For a study of the feminist movement at the turn of the twentieth century, see Christine Bolt, *Sisterhood Questioned? Race, Class, and Internationalism in the American and British Women's Movements, c. 1880s–1970s* (New York: Routledge, 2004), ch. 1.

50. For an analysis of this categorical challenge, see David L. Hull, "Darwinism as a Historical Entity: A Historiographic Proposal," in *The Darwinian Heritage*, ed. David Kohn (Princeton: Princeton University Press, 1985): 774–810.

51. Although the term "feminist" was not frequently used in the United States until the 1910s (and those women studied in this book who lived to see the 1910s did not always consider themselves "feminist"—especially Charlotte Perkins Gilman who preferred to be known as a "humanist"), I use the word in this book according to the modern understanding of the term and, occasionally, to distinguish between those women who were mainly concerned with the vote (suffragists) and those who also critiqued patriarchy, organized religion, marriage, and the family (feminists).

52. Historian of science Bernard Lightman has written extensively on the popularization of Darwin, with a particular emphasis on Britain. See, for example, Bernard Lightman, *Victorian Popularizers of Science: Designing Nature for a New Audience* (Chicago: University of Chicago Press, 2007); Bernard Lightman, ed. *Victorian Science in Context* (Chicago: University of Chicago Press, 1997), especially Bernard Lightman "'The Voices of Nature': Popularizing Victorian Science," 187–211; and Bernard Lightman, "Darwin and the Popularization of Evolution," *Notes and Records of the Royal Society* 64 (March 2010): 5–24.

53. Although, as historian Joan W. Scott and others have argued, "experience" can be a tricky concept on which to build a movement or a history. Joan W. Scott, "The Evidence of Experience," *Critical Inquiry* 17 (Summer 1991): 773–97.

54. Those scholars who have focused on the antifeminist applications of nineteenth-century science were listed in footnote 2. Among these earlier works on women, gender, and science, Cynthia Eagle Russett paid the most attention to how women responded to scientific antifeminism. More recently, historians, especially historians of medicine, have written on the positive uses of science by women in the nineteenth century. See, for example, Carla Bittel, *Mary Putnam Jacobi and the Politics of Medicine in Nineteenth-Century America*, Studies in Social Medicine Series, ed. Allan M. Brandt and Larry Churchill (Chapel Hill: University of North Carolina Press, 2009); Nancy Theriot, "Women's Voices in Nineteenth-Century Medical Discourse: A Step Toward Deconstructing Science," *Signs* 19 (1993), 1–31; Susan Wells, *Out of the Dead House: Nineteenth-century Women Physicians and the Writing of Medicine* (Madison: University of Wisconsin Press, 2001); Regina Morantz-Sanchez, *Conduct Unbecoming: Women Physicians in American Medicine* (Chapel Hill: University of North Carolina Press, 1985); and Arleen Tuchman, *Science Has No Sex: The Life of Marie Zakrzewska, M.D.* (Chapel Hill: University of North Carolina Press, 2006). The scholarship on women, gender, and science is far too expansive to list here; this is a partial listing of works that have particularly influenced this project. Anne Fausto-Sterling, *Myths of Gender: Biological Theories about Women and Men*, rev. ed. (New York: Basic Books, 1992); Patricia Adair Gowaty, ed., *Feminism and Evolutionary Biology: Boundaries, In-*

*tersections, and Frontiers* (New York: Chapman and Hall, 1997); Patricia Adair Gowaty, "Sexual Natures: How Feminism Changed Evolutionary Biology," *Signs* 28 (Spring 2003): 901–93; Robyn Wiegman, *American Anatomies: Theorizing Race and Gender*, New Americanists Series, ed. Donald Pease (Durham: Duke University Press, 1995); Ruth Bleier, ed., *Feminist Approaches to Science* (New York: Pergamon Press, 1986); Janet Sayers, *Biological Politics: Feminist and Anti-Feminist Perspectives* (New York: Tavistock Publications, 1982); and Sally Gregory Kohlstedt, ed. *History of Women in the Sciences: Readings from Isis* (Chicago: University of Chicago Press, 1999).

55. Evelyn Fox Keller, "Feminism and Science," ch. 2 in Evelyn Fox Keller and Helen E. Longino, ed. *Feminism and Science*, Oxford Readings in Feminism (New York: Oxford University Press, 1996), 39. The philosopher of science Sandra Harding makes a similar argument in *The Science Question in Feminism* (Ithaca, NY: Cornell University Press, 1986): "in order to understand the changes occurring in late nineteenth and early twentieth-century science, evidently we need a fuller understanding than intellectual historians and histories of men's worlds can provide. We need to be able to see the events reported by these mainstream histories within histories of gender and sexuality" (66). For additional analyses of feminist science and epistemologies, see also Sally Gregory Kohlstedt and Helen Longino, "The Women, Gender, and Science Question: What Do Research on Women in Science and Research on Gender and Science Have to Do with Each Other?" *Women, Gender, and Science: New Directions*, Special Issue, *Osiris* 12 (1997): 3–15.

56. Elizabeth Grosz, *The Nick of Time: Politics, Evolution, and the Untimely* (Durham, NC: Duke University Press, 2004), 67.

57. Grosz, *The Nick of Time*, 71–3.

58. David N. Livingstone explores the long history of the idea that there were "pre-Adamic" people, especially how pre-Adamic races were often invoked in support of polygenesis, in *Adam's Ancestors: Race, Religion, and the Politics of Human Origins*, Medicine, Science, and Religion in Historical Context Series, ed. Ronald Numbers (Baltimore: Johns Hopkins University Press, 2008).

59. Darwin explains this process in chapter 7, "On the Races of Man," *The Descent of Man* , 194–230.

60. Two studies of race, gender, and evolutionary theory in the nineteenth century have particularly informed my thinking on this topic: Louise Michele Newman, *White Women's Rights: The Racial Origins of Feminism in the United States* (New York: Oxford University Press, 1999); and Gail Bederman, *Manliness and Civilization: A Cultural History of Gender and Race in the United States, 1880–1917*, Women in Culture and Society Series, ed. Catharine R. Stimpson (Chicago: University of Chicago Press, 1995).

61. As Weinbaum writes, "For so long as an unself-reflective portrait of Gilman remains dominant, Gilman's disturbing ideals will continue to haunt feminist self-conception and promulgate the mistaken belief that it is possible for a feminism that does not account for the racialization of gender and sexual formations to be truly liberatory." Alys Eve Weinbaum, "Writing Feminist Genealogy: Charlotte Perkins Gilman and the Reproduction of Racial Nationalism," ch. 2 in *Wayward Reproductions: Genealogies of Race and Nation in Transatlantic Modern Thought* (Durham:

Duke University Press, 2004), 64–5. For additional analyses of the role of evolutionary theory in Charlotte Perkins Gilman's thought, see Carl Degler's classic introduction to the 1966 edition of *Women and Economics*, as well as Carl Degler, "Charlotte Perkins Gilman on the Theory and Practice of Feminism," *American Quarterly* 8 (Spring 1956): 21–39; Gail Bederman, "Not to Sex—but to Race!" Charlotte Perkins Gilman, Civilized Anglo-Saxon Womanhood, and the Return of the Primitive Rapist," in *Manliness and Civilization*, 121–69; Maureen L. Egan, "Evolutionary Theory in the Social Philosophy of Charlotte Perkins Gilman," *Hypatia* 4 (Spring 1989): 102–19; Bernice L. Hausman, "Sex before Gender: Charlotte Perkins Gilman and the Evolutionary Paradigm of Utopia," *Feminist Studies* 24 (Fall 1998): 489–510. For an analysis of Gilman's thought in relation to Spencer's, see Lois N. Magner, "Darwinism and the Woman Question: The Evolving Views of Charlotte Perkins Gilman," in *Critical Essays on Charlotte Perkins Gilman*, ed. Joanne Karpinski, Critical Essays on American Literature Series, ed. James Nagel (New York: G.K. Hall, 1992): 115–28.

62. According to Judith Allen's 2009 intellectual biography of Gilman, the most comprehensive to date, Gilman (the niece of legendary abolitionist Harriet Beecher Stowe) attributed racial differences to cultural and environmental factors (not biology) and "denounced Jim Crow as an immoral response to [chattel slavery's] consequences and aftermath," as well as supported other causes that were considered antiracist in her time. Judith Allen, *The Feminism of Charlotte Perkins Gilman: Sexualities, Histories, Progressivism* (Chicago: University of Chicago Press, 2009), 346–7. Ann D. Gordon contextualizes Stanton's comments about race in "Stanton and the Right to Vote: On Account of Race or Sex," ch. 7 in *Elizabeth Cady Stanton, Feminist as Thinker: A Reader in Documents and Essays*, ed. Ellen Carol DuBois and Richard Cándida Smith (New York: New York University Press, 2007): 111–27. For another excellent study of Stanton's views on race in this same volume, see Michele Mitchell, "'Lower Orders,' Racial Hierarchies, and Rights Rhetoric: Evolutionary Echoes in Elizabeth Cady Stanton's Thought during the Late 1860s," ch. 8 in *Feminist as Thinker*, 128–51.

63. Allen, *The Feminism of . . .* , 344. Allen provides a comprehensive overview of the scholarship regarding race and Gilman's thinking on pages 335–42.

64. For African American men's responses to evolutionary theory, see Eric D. Anderson, "Black Responses to Darwinism, 1859–1915," in *Disseminating Darwinism*, 247–66. Interestingly, Anderson finds that African Americans did not frequently publish responses to evolutionary theory—largely owing to the fact that they had more pressing problems to attend to during this time period—and he contends that there is no evidence that they believed Darwin's ideas were linked to the racism of their time. To the contrary, Anderson argues, "If the chief victims of racism did not interpret Darwin's theory of evolution as the primary source of their oppression, perhaps historians should reconsider certain convenient, time-honored generalizations about evolution, race, and society" (262). Furthermore, when they did cite Darwin directly, African American leaders tended to do so in positive ways, for example, as evidence of monogenesis. For African American responses to the Scopes trial, see, Jeffrey P. Moran, "The Scopes Trial and Southern Fundamentalism in Black and White: Race, Region, and Religion," *Journal of Southern History* 70 (February 2004): 95–120; and Jeffrey P. Moran,

"Reading Race into the Scopes Trial: African American Elites, Science, and Fundamentalism," *Journal of American History* 90 (December 2003): 891–911.

CHAPTER 1

1. The first chapter of Genesis explains that men and women were created simultaneously: "So God created man in his own image, in the image of God created he him; male and female created he them." Gn 1:27. But this passage was quoted much less frequently and, generally, only by women hoping to counter the much better-known rib story, which is Genesis 2:20–22.

2. King James Bible, Gn 3:16.

3. References to Adam and Eve pervaded popular songs, cartoons, and periodicals throughout the nineteenth century. To give just a few examples, the following sources all attempted to tease out in both serious and humorous ways what exactly Adam and Eve meant for modern men and women: A Caithnesian, "On Women," *Boston Weekly Magazine*, July 21, 1804, p. 153. This article reinterpreted the Creation and Fall from Eve's point of view to suggest in fact the blame should rest with Adam, and it was widely reprinted as late as 1818; "On Women," *Weekly Visitor and Ladies' Museum*, May 16, 1818, p. 40; J.E., "The Fall of Man," *Philadelphia Recorder*, October 2, 1824, p. 264; Mada, "Adam and Eve," *New-York Mirror: A Weekly Gazette of Literature and the Fine Arts*, March 6, 1824, p. 253; A Lay Preacher, "A Word for Women," *Home Journal*, October 28, 1848, p. 1; W.L. Tiffany, "Woman," *Godey's Lady's Book* 45 (August 1852): 181; George H. Coomer, "Adam and Eve" [a song], *Flag of Our Union*, October 14, 1865, p. 649; "Adam and Eve," *Circular*, March 5, 1866, p. 408; and two cartoons, "Adam and Eve: Bound to Have That Apple," and "Adam and Eve: The Fall," *Harper's Bazaar*, September 15, 1877, p. 592.

4. E. Courier, "Paintings of Dubufe and Giraud," *New Yorker*, December 8, 1838, p. 180. The paintings were exhibited in Boston, New York, Philadelphia, Baltimore, New Orleans, and Charleston between 1832 and 1835, according to Didier Rykner, *"Adam and Eve* and *Paradise Lost* by Claude-Marie Dubufe Acquired by Nantes," *Art Tribune*, October 22, 2008, http://www.thearttribune.com/Adam-and-Eve-and-Paradise -Lost-by.html.

5. "Adam and Eve," *New York Observer and Chronicle*, June 22, 1833, p. 100.

6. "The Temptation and the Expulsion of Adam and Eve, by Dubufe," *North American Magazine*, June 1833, p. 135.

7. Susan Goodman and Carl Dawson, *William Dean Howells: A Writer's Life* (Los Angeles: University of California Press, 2005), 14.

8. Mark Twain, *The Diaries of Adam and Eve* (first published separately as *Extracts from Adam's Diary* [1904] and *Eve's Diary* [1906]) (New York: Prometheus Books, 2000).

9. Mark Twain, "A Monument to Adam," *San Jose Mercury News*, July 30, 1905, p. 20, reprinted from *Harper's Weekly*.

10. E. Anthony Rotundo, *American Manhood: Transformations in Masculinity from the Revolution to the Modern Era* (New York: Basic Books, 1993), 134.

11. Rotundo, *American Manhood*, 104–5.

12. Rotundo, *American Manhood*, 133.

13. "Pastoral Letter of the General Association of the Massachusetts (Orthodox) to the Churches under Their Care," *Liberator* (Boston), August 11, 1837, excerpts reprinted in Aileen S. Kraditor, *Up from the Pedestal: Selected Writings in the History of American Feminism* (New York: Harper Collins, 1968), 50–52.

14. Elizabeth Cady Stanton, Susan B. Anthony, and Matilda Joslyn Gage, ed. *History of Woman Suffrage*, vol. 1 (Rochester, NY: Susan B. Anthony), 383.

15. Nancy Isenberg, *Sex and Citizenship in Antebellum America*, Gender and American Culture Series, ed. Linda K. Kerber and Nell Irvin Painter (Chapel Hill: University of North Carolina Press, 1998), 71. For an additional history of antebellum feminist thought and women's rights activity focused on New York, see Lori Ginzberg, *Untidy Origins: A Story of Woman's Rights in Antebellum New York* (Chapel Hill: University of North Carolina Press, 2005).

16. "Upshur on Minorities and Majorities," in *Democracy, Liberty, and Property: The State Constitutional Conventions of the 1820's*, ed. Merrill D. Peterson, The American Heritage Series, ed. Leonard W. Levy and Alfred Young (Indianapolis: Bobbs-Merrill Company, 1966), 314. Thanks to Shirley Thompson for sharing this reference with me.

17. Review of Dana's speech, "Woman's Rights," *Lily*, April 1851, p. 28; quoted in Isenberg, *Sex and Citizenship*, 71, fn. 124.

18. Isenberg, *Sex and Citizenship*, 66. Isenberg draws on and quotes from John Quincy Adams, *The Social Compact, Exemplified in the Constitution of the Commonwealth of Massachusetts; with Remarks on the Theories of Divine Right of Hobbes and of Filmer, and the Counter Theories of Sidney, Locke, Montesquieu, and Rousseau, Concerning the Origin of Government: a Lecture, Delivered before the Franklin Lyceum, at Providence, R.I., November 25, 1842* (Providence, R.I.: Knowles and Vose, 1842).

19. Judith Sargent Murray, "On the Equality of the Sexes" (1790) reprinted in *Selected Writings of Judith Sargent Murray*, ed. Sharon M. Harris (New York: Oxford University Press, 1995), 13.

20. Sarah Grimké, "Letters on the Equality of the Sexes, and the Condition of Woman" (1837–1838), reprinted in *Roots of American Feminist Thought*, ed. James L. Cooper and Sheila M. Cooper (Boston: Allyn and Bacon, 1973), 65.

21. Isenberg, *Sex and Citizenship*, 71.

22. Orestes A. Brownson, "The Woman Question," Article II in volume 18 of *The Works of Orestes A. Brownson*, ed. Henry F. Brownson (Detroit, 1885), 403; reprinted in Kraditor, *Up from the Pedestal*, 193.

23. This statement was recorded from a women's rights convention in 1869 and reprinted in the women's press. P.[Parker] P. [Pillsbury], "Woman in Genesis," *Revolution*, November 25, 1869, p. 330.

24. Doctor, "The Fall of Man," *Woodhull and Claflin's Weekly*, November 25, 1871.

25. Editor's Table, "Invention and Intuition," *Godey's Lady's Book and Magazine*, January 1872, p. 93.

26. Rev. John Todd, "Woman's Rights" pamphlet (Boston: Lee and Shepard, 1867),

5. Women's Rights Collection, box 1, folder 3, Sophia Smith Collection, Smith College, Northampton, Massachusetts (hereafter SSC).

27. Jon H. Roberts, *Darwinism and the Divine in America: Protestant Intellectuals and Organic Evolution, 1859–1900*, History of American Thought and Culture Series, ed. Paul S. Boyer, (Madison: University of Wisconsin Press, 1988).

28. Jon H. Roberts, *Darwinism and the Divine in America*, 107. This argument is elaborated in Roberts, "Darwinism, American Protestant Thinkers, and the Puzzle of Motivation," in *Disseminating Darwinism: The Role of Place, Race, Religion, and Gender*, ed. Ronald L. Numbers and John Stenhouse (New York: Cambridge University Press, 1999): 145–72.

29. Roberts, *Darwinism and the Divine in America*, 91–93.

30. Roberts, *Darwinism and the Divine in America*, 103–4.

31. Helen Lefkowitz Horowitz, *Rereading Sex: Battles over Sexual Knowledge and Suppression in Nineteenth-Century America* (New York: Alfred A. Knopf, 2002), 359.

32. Gail Bederman describes "muscular Christianity" in "'The Women Have Had Charge of the Church Work Long Enough': The Men and Religion Forward Movement of 1911–1912 and the Masculinization of Middle-Class Protestantism," *American Quarterly* 41 (September 1989): 432–65. For an extended study of muscular Christianity, see Clifford Putney, *Muscular Christianity: Manhood and Sports in Protestant America, 1880–1920* (Cambridge: Harvard University Press, 2001). See also David Morgan, *Visual Piety: A History and Theory of Popular Religious Images* (Berkeley: University of California Press, 1998), esp. ch. 3, "The Masculinity of Christ," which describes how depictions of Christ changed in the nineteenth-century to reflect new, more manly ideals of men.

33. Kathi Kern, *Mrs. Stanton's Bible* (Ithaca, NY: Cornell University Press, 2001), 79.

34. Philip Schaff, ed., *Popular Commentary on the New Testament by English and American Scholars of Various Evangelical Denominations*, vol. 3 (New York: Scribners, 1882), 219. Quoted in Kern, *Mrs. Stanton's Bible*, 79, fn. 120.

35. Kern references numerous biblical commentaries, all of which emphasized the weakness and venality of Eve. *Mrs. Stanton's Bible*, 78–83, fn. 120–35. Matthew Henry claimed women were "Naturally subordinate" in *The Comprehensive Commentary on the Holy Bible*, ed. William Jenks, 5 vols. (Philadelphia: American Publishing House, 1891), 1:37; quoted in Kern, *Mrs. Stanton's Bible*, 80, fn. 125. See also, Rev. Henry Dexter, *Common Sense as to Woman Suffrage* (Boston: W.L. Greene and Co., 1885), copy residing in the Suffrage Collection, box 3, folder 5, SSC; and A. [Adeline] D. T. W. [Whitney], "The Law of Woman-Life," pamphlet, n.d., Women's Rights Collection, box 1, folder 3, SSC.

36. For studies of masculinity at the end of the nineteenth century, see, for example, Gail Bederman, *Manliness and Civilization: A Cultural History of Gender and Race in the United States, 1880–1917*, Women in Culture and Society Series, ed. Catharine R. Stimpson (Chicago: University of Chicago Press, 1995); Michael S. Kimmel, *Manhood in America: A Cultural History*, 3d ed. (New York: Oxford University Press, 2011), and *The History of Men: Essays in the History of American and British Mascu-*

*linities* (Albany: State University of New York Press, 2005); John F. Kasson, *Houdini, Tarzan, and the Perfect Man: The White Male Body and the Challenge of Modernity in America* (New York: Hill and Wang, 2001); and John Pettegrew, *Brutes in Suits: Male Sensibility in America, 1890–1920*, Gender Relations in the American Experience Series, ed. Joan F. Cashin and Ronald G. Walters (Baltimore: Johns Hopkins University Press, 2007).

37. Bederman provides a thorough and engaging analysis of Roosevelt's transformation from "weakling" to Rough Rider to president in *Manliness and Civilization*, ch. 5, "Theodore Roosevelt: Manhood, Nation, and 'Civilization.'"

38. Grover Cleveland, "Woman's Mission and Woman's Clubs," *Ladies' Home Journal* 22 (May 1905), 3. Quoted in Aileen S. Kraditor, *The Ideas of the Woman Suffrage Movement, 1890–1920* (New York: Columbia University Press, 1965), 16, fn. 4.

39. Claire, "Woman and Evolution" *Woman's Journal*, December 18, 1875, p. 403.

40. T. [Thomas] W. [Wentworth] H. [Higginson], "The Two Camps," *Woman's Journal*, August 19, 1882, p. 257. Even though he held the "utmost of admiration for their great teachings in other ways," Higginson believed that on the question of women's intellect Huxley and Darwin were "open to the suspicion of narrowness" for essentially describing women as lesser men. Higginson, "Darwin, Huxley, and Buckle," *Common Sense about Women* (Boston: Lee and Shephard, 1881), 11.

41. Helen H. Gardener, *Men, Women, and Gods, and Other Lectures* (New York: Truth Seeker Company, 1885), 34, 24.

42. "Woman Suffrage at Newport," reprinted from the *New York Tribune* in the *Revolution*, September 2, 1869, pp. 132–3.

43. Catherine Waugh McCulloch, "The Bible on Women Voting," n.d., no publication information, p. 4, Suffrage Collection, series 1, box 3, folder 6, SSC.

44. Emily Oliver Gibbes, *The Origin of Sin and Dotted Words in the Hebrew Bible* (New York: Charles T. Dillingham, 1893), 127–8. History of Women Collection, reel 549, available through the SSC and the Schlesinger Library, Radcliffe Institute, Cambridge, MA (SLRI).

45. Frances Willard, *Woman in the Pulpit* (Boston: D. Lothrop Company, 1888), 90. Quoted in Kern, *Mrs. Stanton's Bible*, 87, fn. 157.

46. Frederic A. Hinckley, "Woman Suffrage in the Light of Evolution," Pamphlet (Providence: J.A. and R.A. Reid, Printers, 1884), 11. This is a reprint of one of Hinckley's sermons from May 5, 1884, copy residing at Harvard University Library.

47. For studies of the many varieties of evolutionary theory in the late nineteenth century, see Peter J. Bowler, *The Eclipse of Darwinism: Anti-Darwinian Evolution Theories in the Decades around 1900* (Baltimore: Johns Hopkins University Press, 1983); and Martin Fichman, *Evolutionary Theory and Victorian Culture*, Control of Nature Series, ed. Morton L. Schagrin, Michael Ruse, and Robert Hollinger (New York: Humanity Books, 2002).

48. Alan Trachtenberg, *The Incorporation of America: Culture and Society in the Gilded Age*, American Century Series, ed. Eric Foner (New York: Hill and Wang, 1982). For overviews of the Gilded Age, see Rebecca Edwards, *New Spirits: Americans in the Gilded Age, 1865–1905* (New York: Oxford University Press, 2006); Jackson Lears, *Rebirth of a Nation: The Making of Modern America, 1877–1920* (New York: Harper Col-

lins, 2009); Sean Dennis Cashman, *America in the Gilded Age: From the Death of Lincoln to the Rise of Theodore Roosevelt,* 3d ed. (New York: New York University Press, 1993); and T.J. Jackson Lears, *No Place of Grace: Antimodernism and the Transformation of American Culture, 1880–1920* (Chicago: University of Chicago Press, 1994).

49. For classic histories of the women's rights and suffrage movements, see Ellen Carol DuBois, *Feminism and Suffrage: The Emergence of an Independent Women's Movement in America, 1848–1869* (Ithaca, NY: Cornell University Press, 1978); Ellen Carol DuBois, *Woman Suffrage and Women's Rights* (New York: New York University Press, 1998); Nancy Cott, *The Grounding of Modern Feminism* (New Haven: Yale University Press, 1989); Eleanor Flexnor, *Century of Struggle: The Woman's Rights Movement in the United States* (Cambridge: Harvard University Press, 1996); and Aileen S. Kraditor, *The Ideas of the Woman Suffrage Movement, 1890–1920,* reprint of 1965 edition (New York: Norton, 1981).

50. Membership numbers cited in Kern, *Mrs. Stanton's Bible,* 121, fn. 98, from Ruth Bordin, *Women and Temperance: The Quest for Power and Liberty, 1873–1900* (Philadelphia: Temple University Press, 1984), 41.

51. Kern, *Mrs. Stanton's Bible,* 121–3.

52. Lears, *Rebirth of a Nation,* 236–237.

53. Women's rights were not on the agenda for most male social Darwinists. To the contrary, most social Darwinists believed it imperative that women remain in their traditional sphere and bear as many children as possible. For an analysis of Spencer's views on women, see Nancy Paxton, *George Eliot and Herbert Spencer: Feminism, Evolutionism, and the Reconstruction of Gender* (Princeton: Princeton University Press, 1991).

54. For histories of social Darwinism in America, see Richard Hofstadter's classic *Social Darwinism in American Thought* (Philadelphia: University of Pennsylvania Press, 1944); Robert C. Bannister, *Social Darwinism: Science and Myth in Anglo-American Social Thought,* American Civilization Series, ed. Allen F. Davis (Philadelphia: Temple University Press, 1979); Carl N. Degler, *In Search of Human Nature: The Decline and Revival of Darwinism in American Social Thought* (New York: Oxford University Press, 1991); Howard L. Kaye, *The Social Meaning of Modern Biology: From Social Darwinism to Sociobiology* (New Haven: Yale University Press, 1986); and Barry Werth, *Banquet at Delmonico's: Great Minds, the Gilded Age, and the Triumph of Evolution in America* (New York: Random House, 2009).

55. Bannister, *Social Darwinism,* 9.

56. Kathi Kern analyzes women reformers varying interpretations of science in Mrs. Stanton's Bible, 106–9. In addition to the emphasis on social evolution, Kern notes that "While New Thought advocates and Christian Scientists believed that 'true' science was spiritually based, Freethought suffragists placed their faith in the materiality of science and commandeered it as a useful weapon in the campaign for women's equality" (106).

57. Carrie Chapman Catt, "Evolution and Woman Suffrage," speech delivered May 18, 1893, Congress of Representative Women, Chicago, p. 2. Carrie Chapman Catt Papers, box 4, folder 5, Manuscript and Archives Division, New York Public Library, Astor, Lenox, and Tilden Foundations.

58. Catt, "Evolution," 2, 8–9.

59. Kevin Scott Amidon, "Carrie Chapman Catt and the Evolutionary Politics of Sex and Race, 1885–1940," *Journal of the History of Ideas* 68 (April 2007), 308.

60. Louise Michele Newman, *White Women's Rights: The Racial Origins of Feminism in the United States* (New York: Oxford University Press, 1999).

61. Beryl Satter, *Each Mind a Kingdom: American Women, Sexual Purity, and the New Thought Movement, 1875–1920* (Los Angeles: University of California Press, 1999), 185. For extended analyses of educated suffrage, see Kern, *Mrs. Stanton's Bible*, 106–16; and Ann D. Gordon "Stanton and the Right to Vote: On Account of Race or Sex," ch. 7 in *Elizabeth Cady Stanton, Feminist as Thinker: A Reader in Documents and Essays*, ed. Ellen Carol DuBois and Richard Cándida Smith (New York: New York University Press, 2007): 111–27.

62. Kern, *Mrs. Stanton's Bible*, 112–4.

63. Ann D. Gordon, "Stanton and the Right to Vote," 111.

64. Gordon, "Stanton and the Right to Vote," 123.

65. Michele Mitchell, "'Lower Orders,' Racial Hierarchies, and Rights Rhetoric: Evolutionary Echoes in Elizabeth Cady Stanton's Thought during the Late 1860s," in *Elizabeth Cady Stanton, Feminist as Thinker*, 129. For a history of the relationship between the abolition and women's rights movements including a new, more nuanced interpretation of debates over the Fourteenth and Fifteenth Amendments, see Faye E. Dudden, *Fighting Chance: The Struggle over Woman Suffrage and Black Suffrage in Reconstruction America* (New York: Oxford University Press, 2011).

66. Flavia Alaya, for example, argues that nineteenth-century science "not only strengthened the opposition to feminism but disengaged the ideals of feminists themselves from their philosophical [egalitarian] roots." Alaya, "Victorian Science and the 'Genius' of Woman," *Journal of the History of Ideas* 38 (1977): 262. Similarly, Fiona Erksine contends that "it was this endorsement of sexual difference, this complicity in the social construct of Victorian patriarchy that rendered the Victorian women's movement so vulnerable to the attack of the scientific anti-feminists." Fiona Erskine, "The *Origin of Species* and the Science of Female Inferiority," in *Charles Darwin's The Origin of Species: New Interdisciplinary Essays*, ed. David Amigoni and Jeff Wallace, Texts in Culture Series, ed. Stephen Copley and Jeff Wallace (New York: Manchester University Press, 1995), 115. For related arguments in this vein, see Aileen S. Kraditor, *The Ideas of the Woman Suffrage Movement, 1890–1920*, Anchor Books edition, (Garden City: Doubleday, 1971), 15–17, 39; Rosalind Rosenberg, *Beyond Separate Spheres*, 14–15; and Rosalind Rosenberg, "In Search of Woman's Nature, 1850–1920," *Feminist Studies* 3 (Fall 1975): 141–54.

67. Ann Gordon, "Stanton and the Right to Vote," 113.

68. E.C.S., "Miss Becker on the Difference in Sex," *Revolution*, September 24, 1868, pp. 184–5. For an additional analysis of the turn away from egalitarian arguments and the role of evolutionary theory in Stanton's political thought, see Sue Davis, *The Political Thought of Elizabeth Cady Stanton: Women's Rights and the American Political Traditions* (New York: New York University Press, 2008).

69. Thomas Laqueur, "Orgasm, Generation and the Politics of Reproductive Biology," reprinted in *The Making of the Modern Body: Sexuality and Society in the Nine-*

*teenth Century,* ed. Thomas Laqueur and Catherine Gallagher (Berkeley: University of California Press, 1987). See also Thomas Laqueur, *Making Sex: Body and Gender from the Greeks to Freud* (Cambridge: Harvard University Press, 1990); for another canonical study of the ramifications of Enlightenment philosophy for women, see Carole Pateman, *The Sexual Contract* (Stanford: Stanford University Press, 1988).

70. Darwin, *Descent of Man,* 191–2.

71. In *Each Mind a Kingdom,* Beryl Satter groups Frances Willard and Stanton together under the banner of "reform Darwinist" (11). While it is true that both women suggested that evolution and women's rights went hand-in-hand, my research suggests that what they meant by evolution was very different and often contrasting, especially with regard to what evolution meant for religious orthodoxy. While Willard disagreed with male social Darwinists regarding women's place in the universe, she agreed with them regarding the progressive course of evolution and its compatibility with Christian ascendency. Stanton rejected the NAWSA merger, detested the way that the WCTU women and their tactics had infiltrated the women's rights movement, and latched on to evolutionary science largely because she believed it discredited the Genesis creation story and orthodox Christianity. For more on Stanton's views regarding the merger and the WCTU, see Kern, *Mrs. Stanton's Bible,* 121–9.

72. The *Revolution,* which Stanton edited, and later the *Woman's Tribune,* edited by Stanton ally Clara Bewick Colby, regularly ran articles critiquing the Adam and Eve story and offering alternatives. An 1868 article in the *Revolution* lamented that "infanticide, idiocy, prostitution, public and in wedlock, and infamous human degradation are but the outgrowths, the wormwood results of this malicious, libelous doctrine that woman was created inferior, subject to masculine will and dictation." Thomas W. Organ, M.D., "Woman Wronged," *Revolution,* April 9, 1868, p. 214. Throughout 1869, the *Revolution* ran regular articles critiquing the Genesis creation story and offering alternatives from science and other world religions. For several issues, the *Woman's Tribune* also ran accounts of various religions' creation stories. In the January 1887 edition of the *Woman's Tribune,* for example, two side-by-side articles address the Eve issue: E.T. Grover, "Woman and the Curse," and Lucie L. Prudhomme "Milton's Eve." Grover argued that "upon that supposed Divine curse man has grounded all his opinions, theory and practice in regard to his relations to and rights and power over, this cursed half of the human family." Prudhomme critiqued Milton's depiction of Eve in "Paradise Lost," concluding that woman fell first "not because she was the weaker, but because she was the stronger, requiring greater attention, mightier temptation, Satan himself for the reduction of her strength."

73. Elizabeth Cady Stanton, *Woman's Bible, parts I and II,* reprinted through the American Women: Images and Realities Series (New York: Arno Press, 1972), 20. Page citations are to this edition. The *Woman's Bible* was initially published in two parts; volume 1 in 1895 and volume 2 in 1898.

74. Matilda Joslyn Gage, *Woman, Church, and State: The Original Exposé of Male Collaboration Against the Female Sex* (Chicago: C.H. Kerr, 1893, reprint Watertown: Persephone Press, 1980), 12. Page citations are to the reprint. In addition to Gage's *Woman, Church and State,* other women advanced similar arguments about the impor-

tance of the biblical creation story for women's oppression, see, for example, Gibbes, *The Origin of Sin*; and Eliza Burt Gamble, *The God-Idea of the Ancients, or Sex in Religion* (New York: Putnam, 1897).

75. Gage, *Woman, Church and State*, 235.

76. Matilda Josyln Gage to Thomas Clarkson Gage, March 7, 1890, Gage Papers, SLRI; quoted in Sally Roesch Wagner, *Matilda Joslyn Gage: She Who Holds Up the Sky*, 2d ed. (Aberdeen, SD: Sky Carrier Press, 2002), 57, fn. 72. For more on Gage in relation to suffrage history, see Leila R. Brammer, *Excluded from Suffrage History: Matilda Joslyn Gage, Nineteenth-Century American Feminist*, Contributions in Women's Studies, no. 182 (Westport, CT: Greenwood Press, 2000).

77. Stanton led an unsuccessful attempt to remove the word "Christian" from the mission statement of the Women's Loyal National League during the Civil War. *History of Woman Suffrage*, vol. 2, 891; Elizabeth Cady Stanton to William Lloyd Garrison, Jr., January 6, [1896], film, reel 35. Quoted in Lori Ginzberg, *Elizabeth Cady Stanton: An American Life* (New York: Hill and Wang, 2009), 172, fn. 52.

78. After the collapse of the *Revolution*, the *Woman's Tribune* became the unofficial organ of the NWSA, while the *Woman's Journal* remained the newspaper of record for the AWSA and, later, the united NAWSA. The *Woman's Tribune* was very much influenced by Stanton and reflected her broad interests; the *Woman's Journal* mainly printed articles pertaining to suffrage. For a detailed history of the women's rights press, see Martha M. Solomon, ed., *A Voice of Their Own: The Woman Suffrage Press, 1840–1910* (Tuscaloosa: University of Alabama Press, 1991).

79. For example, the 1885 resolution is reprinted in Stanton's autobiography, *Eighty Years and More* (London: T. Fisher Unwin, 1898), 381. Minutes from one meeting where such a resolution was debated were reprinted in the *Woman's Tribune*, March 1885. Kern describes Stanton's anticlericalism and these resolutions in, *Mrs. Stanton's Bible*, 95–98; see also Lois Banner, *Elizabeth Cady Stanton: A Radical for Women's Rights*, Library of American Biography Series, ed. Oscar Handlin (Boston: Little Brown, 1980), 154–5.

80. Elizabeth Cady Stanton, "Woman and the Church," *Lucifer the Light-Bearer*, July 17, 291 (this newspaper did not abide by the Christian calendar, so 291 corresponds to 1891). SSC.

81. Susan Jacoby, *Freethinkers: A History of American Secularism* (New York: Metropolitan Books, 2004), especially chapters 5 and 6 on the role of evolution in freethought and on the Golden Age of freethought in the United States, respectively. Jacoby discusses the role of freethought in the antebellum women's rights movement (ch. 3) and in the late nineteenth century, 194–205. Jacoby argues that because of its heretical basis, the freethought movement and its adherents have largely been written out of American history, a claim that definitely holds true among freethinking feminists. For an anthology of work by freethinking women, see Annie Laurie Gaylor, ed., *Women without Superstition "No Gods—No Masters": The Collected Writings of Women Freethinkers of the Nineteenth and Twentieth Centuries* (Madison, WI: Freedom from Religion Foundation, 1997).

82. Jacoby, *Freethinkers*, 144.

83. Ginzberg, *Elizabeth Cady Stanton*, 171–2.

84. Kern, *Mrs. Stanton's Bible*, 64.

85. Kathi Kern, "'Free Woman Is a Divine Being, the Savior of Mankind': Stanton's Exploration of Religion and Gender," in *Elizabeth Cady Stanton, Feminist as Thinker: A Reader in Documents and Essays*, ed. Ellen Carol DuBois and Richard Cándida Smith (New York: New York University Press, 2007), 102. Although, as Kern establishes, Stanton's involvement in freethought also "undercut her commitment to collective female agency" (94) as many leading freethinkers "offered a very narrow conception of womanhood" (103).

86. Stanton, "Woman and the Church."

87. For an analysis of Stanton's time in England and France in the early 1880s, see Kern, *Mrs. Stanton's Bible*, 50–71. For Banner's take on Stanton's time abroad, see Banner, *Elizabeth Cady Stanton*, 166–7.

88. Theodore Stanton and Harriot Stanton Blatch, ed., *Elizabeth Cady Stanton As Revealed in Her Letters, Diary, and Reminiscences*, vol. 2 (New York: Harper and Brothers, 1922), 198. Diary entry is from November 25, 1882.

89. Stanton to Elizabeth Smith Miller, March 5, 1887, Theodore Stanton Collection, Elizabeth Cady Stanton Papers, Rutgers University Libraries, New Brunswick, New Jersey; quoted in Kern, *Mrs. Stanton's Bible*, 69, fn. 69.

90. Elizabeth Cady Stanton, "The Woman's Bible," *Index*, August 19, 1886, p. 87.

91. Stanton, *Woman's Bible*, preface. Kern has written the definitive account of the history, publication, and significance of the *Woman's Bible* with *Mrs. Stanton's Bible*. In introductions to reprints of the *Woman's Bible*, two other scholars have also analyzed it: see Barbara Welter's introduction to *The Original Feminist Attack on the Bible* (New York: Arno, 1974); and Maureen Fitzgerald's foreword to the *Woman's Bible* (Boston: Northeastern University Press, 1993).

92. Stanton, *Woman's Bible* excerpt, *Woman's Tribune*, April 6, 1895; Stanton, *Woman's Bible*, 24.

93. For a comprehensive analysis of the book's reception, see Kern, *Mrs. Stanton's Bible*, 172–222.

94. "The Woman's Bible," *Omaha World Herald*, May 22, 1895, p. 4.

95. E.M. King, "The Bible and Woman's Rights," *Woman's Tribune*, November 18, 1891, p. 311.

96. S.R. Shepherd, "The Vindication of Woman and Sex," *Lucifer the Light-Bearer*, June 28, 1895. SSC.

97. This incident is recounted in detail in Kern, *Mrs. Stanton's Bible*, 182–5.

98. Rachel Foster Avery, ed. Proceedings of the Twenty-Eighth Annual Convention of the National American Woman Suffrage Association, held in Washington, DC, January 23–28, 1896 (Philadelphia: Alfred J. Ferris, 1896), 91. Quoted in Kern, 184, fn. 42.

99. Kern, *Mrs. Stanton's Bible*, 187–9. Kern cites Gilman's reaction, from her diary January 28, 1896, Charlotte Perkins Gilman Papers, SLRI, 187, fn. 51.

100. The NAWSA election of 1900 and the rift between Stanton's followers and Catt and her followers is analyzed in Kern, *Mrs. Stanton's Bible*, 200–206. See also Grace Farrell, *Lillie Devereux Blake: Retracing a Life Erased* (Amherst: University of Massachusetts Press, 2002), 168–72. Stanton's children and core followers were increasingly distressed to see much of the early suffrage movement attributed to Stanton's longtime

collaborator Susan B. Anthony, as evidenced by the decision to name the Nineteenth Amendment the "Susan B. Anthony Amendment," which came as a slap in the face to Stanton's daughter Harriot, according to Lori Ginzberg, *Elizabeth Cady Stanton*, 191. In the years following Stanton's death in 1902, Stanton's children waged a public relations campaign to have Stanton's efforts and writings restored to prominence in the history of the women's rights movement, but for much of the twentieth century her role was diminished by insider histories that privileged the contributions of Anthony and the less outspoken, less controversial women who came to prominence after 1890. Lori Ginzberg describes Stanton's ire at the next generation's celebration of Anthony, not her, on 183–4 and explains how "by the early twentieth century, Anthony was widely celebrated as the sole founder and leader of the cause of woman's rights" (191), despite Stanton's efforts to preserve her legacy. Stanton's foundational role in the women's rights movement was not restored until the emergence of feminist history in the 1970s and the publication of biographies by Lois Banner and Elisabeth Griffith (Ginzberg, *Elizabeth Cady Stanton*, 192).

101. "Evolutionism" refers to the general, nonspecific uses of the term "evolution," generally to mean planned, teleological progress, often grounded in a Spencerian, rather than Darwinian, framework. Michael Ruse, *The Evolution-Creation Struggle* (Cambridge: Harvard University Press, 2005).

CHAPTER 2

1. Helen Hamilton Gardener, Last Will and Testament, Helen Hamilton Gardener Papers in the Women's Rights Collection, M-133, folder 73, Schlesinger Library, Radcliffe Institute for Advanced Study, Harvard University (SLRI).

2. Cora Rigby, "The Diplomatic Corps," written as part of a Woman with a Career series for a newspaper; the original source is not clear from the clip, May 2, 1925. Helen Hamilton Gardener Papers in the Women's Rights Collection, M-133, folder 74, SLRI.

3. Margaret W. Rossiter explains that as the field of science professionalized between 1880 and 1910, it also became increasingly masculine, relegating women to helping positions, such as lab assistant and secretary. Rossiter, *Women Scientists in America: Struggles and Strategies to 1940* (Baltimore: Johns Hopkins University Press, 1982), esp. ch. 4. See also, Rossiter, "Women's Work in Science, 1880–1910," in *History of Women in the Sciences, Readings from Isis*, ed. Sally Gregory Kohlstedt (Chicago: University of Chicago Press, 1999): 287–304. For a study of women's scientific involvement focusing on the first two-thirds of the nineteenth century, see Nina Baym, *American Women of Letters and the Nineteenth-Century Sciences: Styles of Affiliation* (New Brunswick: Rutgers University Press, 2002). For histories of professional science in America, see Sally Gregory Kohlstedt, *The Formation of the American Scientific Community: The American Association for the Advancement of Science, 1848–60* (Urbana: University of Illinois Press, 1976); Sally Gregory Kohlstedt, Michael M. Sokal, and Bruce V. Lewenstein, *The Establishment of Science in America: 150 Years of the American Association for the Advancement of Science* (New Brunswick: Rutgers University Press, 1999); and Philip J. Pauly, *Biologists and the Promise of American Life: From Meriwether Lewis to Alfred Kinsey* (Princeton: Princeton University Press, 2000).

For studies of women and science in England and Europe, see Marina Benjamin, ed., *Science and Sensibility: Gender and Scientific Enquiry, 1780–1945* (Cambridge, UK: Blackwell, 1999); Sally Shuttleworth, Gavin Dawson, and Richard Noakes, "Women, Science, and Culture: Science in the Nineteenth-Century Periodical," *Women: A Cultural Review* 2 (2001): 57–70; Anne B. Shteir, *Cultivating Women, Cultivating Science: Flora's Daughters and Botany in England, 1760 to 1860* (Baltimore: Johns Hopkins University Press, 1996); and Anne B. Shteir, "Elegant Recreations? Configuring Science Writing for Women," in *Victorian Science in Context*, ed. Bernard Lightman (Chicago: University of Chicago Press, 1997): 236–55.

4. Helen H. Gardener, "Sex in Brain," reprinted in *Facts and Fictions of Life*, 3d ed. (Boston: Arena, 1895), 97–8.

5. Antoinette Brown Blackwell, *The Sexes throughout Nature* (New York: G.P. Putnam's Sons, 1875, reprinted through The Pioneers of the Woman's Movement series, Westport: Hyperion Press, 1976). Page citations are to the reprint edition.

6. Most previous scholarship on women and nineteenth-century science has emphasized the ways in which science was used to thwart women's advancement. The best and most thorough account of the sexist uses of science in the nineteenth century remains Cynthia Eagle Russett, *Sexual Science: The Victorian Construction of Womanhood* (Cambridge: Harvard University Press, 1989). Other related sources are cited in the introduction, footnote 2. More recently, scholars including myself have begun to ask, In what ways did women use science for feminist purposes? For more work in this vein, see also Carla Bittel, *Mary Putnam Jacobi and the Politics of Medicine in Nineteenth-Century America*, Studies in Social Medicine Series, ed. Allan M. Brandt and Larry Churchill (Chapel Hill: University of North Carolina Press, 2009).

7. One might also consider Gardener's donation an attempt to give agency to the objects of scientific research, in this case, women. Feminist science scholar Donna Haraway argues that as part of any feminist critique of science, one must "come to terms with the agency of objects." Donna Haraway, "The Science Question in Feminism," in *Feminism and Science, Oxford Readings in Feminism*, ed. Evelyn Fox Keller and Helen E. Longino (New York: Oxford University Press, 1996), 259.

8. The Darwin Correspondence Project, housed at Cambridge University, has a "Gender and Darwin" project that encourages the study of Darwin's correspondence with women. http://www.darwinproject.ac.uk/ See also, Joy Harvey, "Darwin's 'Angels': The Women Correspondents of Charles Darwin," *Intellectual History Review* 19, no. 2 (2009): 197–210. For a study of Darwin's American correspondent Mary Treat, see Tina Gianquitto, "Of Spiders, Ants, and Carnivorous Plants: Domesticity and Darwin in Mary Treat's Home Studies in Nature," in *Coming into Contact: Explorations in Ecocritical Theory and Practice*, ed. Annie Merrill Ingram, Ina Marshall, Daniel J. Philippon, and Adam Sweeting (Athens: University of Georgia Press, 2007): 239–49; and Tina Gianquitto, *"Good Observers of Nature": American Women and the Scientific Study of the Natural World, 1820–1885* (Athens: University of Georgia Press, 2007).

9. Grace Anna Lewis, "Science for Women," paper presented at the third annual Association for the Advancement of Women Congress, held in Syracuse, New York, October 14, 1875, pp. 69–70. Miscellaneous organizations collection, box 10, folder 3, Sophia Smith Collection, Smith College, Northampton, Massachusetts (SSC).

10. See, for example, "A Woman Scientist," *Woman's Tribune*, September 1886, p. 1; C.A.W., "Scientific Work by Women," *Woman's Tribune*, February 1886, p. 2; and "Women in Science," *Woman's Tribune*, August 26, 1893; "Scientific Lecture by Miss Sara S. Hennell," *Woman's Tribune*, June 1, 1889, p. 1; and "Bits of Science," *Woman's Tribune*, February 27, 1892. While the *Woman's Journal* did not publish as many "women in science" news stories as the *Woman's Tribune*, it reported on the Edward Clarke debate in nearly every edition from 1873–1875 and published Blackwell's series of scientific articles that are explored in chapters 2 and 3. Reflecting the schism described in chapter 1, however, only the *Woman's Tribune*, which was closely aligned with Stanton, regularly reported on science in the late 1880s and 1890s.

11. Elizabeth Cazden, *Antoinette Brown Blackwell: A Biography* (Old Westbury, NY: Feminist Press, 1983); Cazden recounts Blackwell's struggles to enter Oberlin University's Theology Department, 35–40. Finney, however, did support Blackwell's right to speak in his classes, which most of her other professors did not, and this marked a turning point in her career at Oberlin.

12. Antoinette Brown Blackwell, *Antoinette Brown Blackwell: The First Woman Minister*, ed. and with an introduction by Mrs. Claude U. Gilson, unpublished autobiography, 169–70. Blackwell Family Papers, M-35, folders 3–14, SLRI. In chapter 5 of her biography of Blackwell, Cazden attributes Blackwell's decision to leave the pulpit to her increasing feelings of isolation and the pressure of having to preach orthodoxy when it conflicted with her more liberal religious beliefs and upbringing.

13. Antoinette Brown Blackwell, ordination records, copy within drafts of Blackwell's unpublished autobiography. Blackwell Family Papers, M-35, folders 3–14, SLRI. In an 1881 letter to fellow pioneering female minister Olympia Brown, Blackwell recalled that "scientific difficulties" led to her crisis of faith and subsequent resignation. Antoinette Brown Blackwell to Olympia Brown, January 10, 1881, p. 5. Olympia Brown Papers, Series III, folder 134, SLRI. In her autobiography she wrote that her spiritual crisis was precipitated by her reading of Darwin, Spencer, and *Popular Science Monthly*. It is important to note, however, that Blackwell's account of her scientific difficulties does not mesh exactly with the chronology. Darwin's *Origins of Species* was not published until 1859, and *Popular Science Monthly* was not in circulation until 1872. In the early 1850s, she did read the early writings of Herbert Spencer, such as *Social Statics* (1851), which first introduced her to the concept of gradual evolution. In this work, Spencer had not yet adopted the antifeminist views that characterized his later writings. Another Blackwell biography links her scientific awakening to Spencer:

> She was fascinated with the new discoveries of present-day scientists. . . . She read and reread portions of both the Old and New Testament, studying and comparing their opposing philosophies and beliefs; pored over the two volumes of the young English writer Herbert Spencer, *Social Statistics* [sic] and *The Development Hypothesis*. Spencer's objection to one's accepting the fact of there being different forms of life, without understanding and explaining how their development took place, appealed strongly to her sense of logic. She thought his insistence that the order of nature was a slow and gradual process seemed reasonable. Laura Kerr, *Lady in the Pulpit* (New York: Woman's Press, 1951), 145.

14. For Darwin's reply to Blackwell (which was addressed "Dear Sir"), see Charles Darwin to Antoinette Brown Blackwell, November 8, no year, Blackwell Family Papers, SLRI. The letter itself is very brief and thanks Blackwell for sending him a copy of *Studies in General Science.*

15. For an example of this later work, see Blackwell, *The Making of the Universe, Evolution the Continuous Process Which Derives the Finite from the Infinite* (Boston: Gorham Press, 1914), in which she argues, for example,

> The claim that Deity manifests himself and His energizing in Nature, is true in the highest and broadest sense; but it is not literally true as a direct statement of fact. He has devised a distinct system of inter-activities, a definite method evolutionary in its inherent ongoing; within its limits all finite activity must confine itself, and for the work done each actor acts for itself does and must accept the outreaching results. (19)

16. Blackwell, *The Sexes,* 234–5.

17. Cazden, *Antoinette Brown Blackwell,* 146.

18. Cazden, *Blackwell,* 188.

19. Blackwell, *The Sexes,* 231–2.

20. Antoinette Brown Blackwell, "The Savans of the Woman Question," *Woman's Journal,* July 22, 1876, p. 1.

21. Blackwell, *The Sexes,* 14, 12.

22. Blackwell, *The Sexes,* 6.

23. Blackwell, *The Sexes,* 22.

24. Blackwell, *The Sexes,* 235.

25. Blackwell, *The Sexes,* 224.

26. "Last Will and Testament of Miss Sophia Smith" (Northampton: Gazette Printing, 1872), 10. Courtesy of the Smith College Archives, Smith College, Northampton, Massachusetts (hereafter SCA).

27. Rev. L. Clark Seelye, Inaugural Address, July 14, 1875 (Springfield: Clark W. Bryan and Company, 1875), 23. Bound in the *Official Circulars Book, 1872–1884,* SCA. Interestingly, Seelye cited Clarke's *Sex in Education* to argue for the importance of single-sex education. See Rev. L. Clark Seelye, "The Need of a Collegiate Education for Women," July 28, 1874 (North Adams: American Institute of Instruction, 1874). Bound in "Official Circulars, 1872–1884," SCA.

28. *Memorial: Alfred Theodore Lilly,* June 7, 1890 (Florence: Trustees of Florence, 1890). Buildings Collection, box 76, SCA.

29. Rev. L. Clarke Seelye, *The Early History of Smith College, 1871–1910* (Boston: Houghton Mifflin, 1923), 65. SCA.

30. Gertrude Gane to "My Dear Mamma," October 27, 1893, Student Letters Collection, SCA.

31. Myra Sampson, Ph.D., Chairman of the Department of Zoology, "Zoology in Smith College," *Smith Alumnae Quarterly* 26 (November 1934), 1, SCA.

32. "Lilly Hall of Science," *Woman's Tribune,* August 1886, p. 2. For an in-depth study of how nineteenth-century women learned about and engaged in evolutionary science at neighboring Mount Holyoke, see Miriam R. Levin, *Defining Women's Scientific*

*Enterprise: Mount Holyoke Faculty and the Rise of American Science* (Hanover: University Press of New England, 2005). For a study of women's science education earlier in the nineteenth century, see Deborah Jean Warner, "Science Education for Women in Antebellum America," in *History of Women in the Sciences: Readings from Isis*, ed. Sally Gregory Kohlstedt (Chicago: University of Chicago Press, 1999): 191–200.

33. "Lilly Hall of Science," *Woman's Journal*, July 3, 1886, p. 1.

34. For a history of women's clubs, including the AAW, see Karen Blair, *The Club-woman as Feminist: True Womanhood Redefined, 1868–1914* (New York: Holmes and Meier, 1980). The AAW is discussed on pages 40–51. According to Blair, the AAW disbanded after the founding of the General Federation of Women's Clubs made its work obsolete. See also Anne Firor Scott, *Natural Allies: Women's Associations in American History* (Urbana: University of Illinois Press, 1991).

35. Anna Garlin Spencer, *The Council Idea and a Tribute to May Wright Sewall*(New Jersey: J. Heidingsfeld Company, 1930), 4; quoted in Blair, *Clubwoman*, 47, n. 53.

36. Maria Mitchell, Presidential Address at the third annual AAW convention, "Papers Read at the Third Congress of Women, Syracuse, NY October 13, 14, 15 1875," 6. Miscellaneous Organizations Collection, box 10, folder 4, SSC. For biographies of Mitchell, see, Sally Gregory Kohlstedt, "Maria Mitchell and the Advancement of Women in Science," in *Uneasy Careers and Intimate Lives, Women in Science, 1789–1979*, ed. Pnina G. Abir-Am and Dorinda Outram, foreword by Margaret Rossiter (New Brunswick: Rutgers University Press, 1987): 129–46; and Renée Bergland, *Maria Mitchell and the Sexing of Science: An Astronomer among the American Romantics* (Boston: Beacon Press, 2008).

37. Maria Mitchell, "The Need of Women in Science," 1876 AAW Annual Congress, 9–10, Miscellaneous Organizations Collection, box 10, folder 4, SSC.

38. Mitchell, Presidential Address.

39. *Popular Science Monthly* (PSM) ran an announcement about the annual meeting of the National Science Club for Women in 1895, which does not appear to be affiliated with the AAW. This club met at what later became the Smithsonian Museum and seems mainly to have attracted women who came to Washington, D.C., for "scientific study and investigation." The group seems to have existed at least since 1891, but the final line of the notice did not bode well for the group's future: "The library needs gifts of books and pamphlets in science." "American Women in Science," PSM 47 (June 1895), 283. For more on Maria Mitchell's efforts to involve women in science, see also Rossiter's discussion of her in *Women Scientists in America*.

40. Maria Mitchell, "Report of Committee on Science," 1875 AAW Congress, Syracuse, New York, accessed via the Gerritsen Collection of Women's History online.

41. This is based on my reading of several AAW programs and pamphlets and tallying the topics by category. The "miscellaneous organization" collection at the SSC has most of the AAW meeting programs, as well as other materials related to this group.

42. Grace Anna Lewis, "Science for Women," pp. 69–70.

43. Daniel Patrick Thurs, *Science Talk: Changing Notions of Science in American Popular Culture* (New Brunswick: Rutgers University Press, 2007), 4. Thurs discusses

the relationship between Darwinian evolutionary theory and the establishment of scientific authority in chapter 2, "Evolution: Struggling over Science."

44. Thurs, *Science Talk*, 77.

45. Thomas F. Gieryn, "Boundary-Work and the Demarcation of Science from Non-Science: Strains and Interests in Professional Ideologies of Scientists," *American Sociological Review* 48 (December 1983): 781–95. See also, Gieryn, *Cultural Boundaries of Science: Credibility on the Line* (Chicago: University of Chicago Press, 1999).

46. Thurs elaborates on the notion of boundary-work, *Science Talk*, 5–7.

47. Thurs, *Science Talk*, 5.

48. Thurs, *Science Talk*, 134–5.

49. Antoinette Brown Blackwell, "Women in Science," *Independent*, May 7, 1891, p. 6.

50. Blackwell, *The Sexes*, 230.

51. As has been mentioned previously, the best and most thorough account of the scientific studies of sex difference that emerged in the late nineteenth century is Cynthia Eagle Russett's *Sexual Science: The Victorian Construction of Womanhood* (Cambridge: Harvard University Press, 1989). While she does mention feminist responses to scientific theories of women's inferiority, the main point of the book is to document various scientific theories of women's inferiority. Rosalind Rosenberg's *Beyond Separate Spheres: Intellectual Roots of Modern Feminism* (New Haven: Yale University Press, 1982) is another excellent study of the Clarke debates. Rosenberg begins with debates about Clarke's *Sex in Education* and then traces the first generation of women social scientists who dismantled biological determinism in the early twentieth century. See also Marie Tedesco, "Science and Feminism: Conceptions of Female Intelligence and Their Effect on American Feminism, 1859–1920" [Ph.D. diss., Georgia State University, 1978]; Carroll Smith-Rosenberg and Charles Rosenberg, "The Female Animal: Medical and Biological Views of Woman and Her Role in Nineteenth-Century America," *Journal of American History* 60 (September 1973): 332–56; Evelleen Richards, "Darwin and the Descent of Woman," in *The Wider Domain of Evolutionary Thought*, ed. David Oldroyd and Ian Langham (Boston: D. Reidel Publishing Company 1983): 57–111; Janice Law Trecker, "Sex, Science and Education," *American Quarterly* 26 (October 1974): 352–66; Susan Sleeth Mosedale, "Science Corrupted: Victorian Biologists Consider 'The Woman Question,'" *Journal of the History of Biology* 11 (Spring 1978): 1–55; John S. Haller, Jr., and Robin M. Haller, *The Physician and Sexuality in Victorian America* (Urbana: University of Illinois Press, 1974).

52. Nancy Woloch, *Women and the American Experience*, 5th ed. (New York: McGraw Hill, 2011), 272. For secondary sources on women going to college in the second half of the nineteenth century, see, for example, Helen Lefkowitz Horowitz, *Alma Mater: Design and Experience in the Women's Colleges from Their Nineteenth-Century Beginnings to the 1930s* (New York: Knopf, 1984); Barbara Miller Solomon, *In the Company of Educated Women: A History of Women and Higher Education in America* (New Haven: Yale University Press, 1985); Lynn D. Gordon, *Gender and Higher Education in the Progressive Era* (New Haven: Yale University Press, 1990).

53. "Biology and 'Woman's Rights,'" reprinted from *Quarterly Journal of Science* in PSM 14 (December 1878): 201–13.

54. Charles Darwin, *The Descent of Man, and Selection in Relation to Sex*, 2d ed. (1879), with an introduction by Adrian Desmond and James Moore (New York: Penguin Classics, 2004), 628–9.

55. George Romanes, "Mental Differences of Men and Women," PSM 31 (July 1887), 392.

56. Romanes, "Mental Differences," 383.

57. Miss M. A. Hardaker, "Science and the Woman Question," PSM 22 (March 1882), 581–2.

58. Herbert Spencer and Darwin agreed on this point. See Herbert Spencer, "Psychology of the Sexes," PSM 4 (November 1873): 30–38. Spencer asked rhetorically, "[A]re the mental natures of men and women the same?" Surely, not, he replied. "That men and women are mentally alike, is as untrue as that they are alike bodily. Just as certainly as they have physical differences which are related to the respective parts they play in the maintenance of the race, so certainly have they psychical differences, similarly related to their respective shares in the rearing and protection of offspring" (31).

59. Darwin, *Descent*, 631.

60. Charles Darwin to Caroline Kennard, January 9, 1881, the Darwin Papers, Cambridge University Library. Quoted in Robert Richards, *Darwin and the Emergence of Evolutionary Theories of Mind and Behavior* (Chicago: University of Chicago Press, 1987), 189, fn. 17.

61. Here, too, Darwin's writings were multivalent. S. Tolver Preston used the "peahen" quote to argue in *Nature* that women should be educated since Darwin had established that men limiting female education have also "enormously injured his own advance—by inheritance." S. Tolver Preston, "Evolution and Female Education," *Nature* 22, September 23, 1880, p. 485.

62. See also, Joseph Le Conte, "The Genesis of Sex," PSM 16 (December 1879): 167–79. According to Le Conte, sex developed as part of "the most universal of all the laws of evolution," the law of differentiation. Sexual differentiation increased as species became more complex. Sexual differentiation also increased "in the higher as compared with the lower races of man" (176). For evolutionary arguments against the idea that men were "naturally" superior and more varied, see David G. Ritchie, *Darwinism and Politics*, 2d ed. (London: Swan Sonnenschein & Co., 1891), 62–75; and Emmet Densmore, *Sex Equality* (New York: Funk and Wagnalls, 1907). Densmore's book is "based on the arguments of Darwin and Spencer" and finds that women are not inherently inferior to men but rather have been limited by cultural factors and are deserving of full equality, including the vote.

63. Edward H. Clarke, *Sex in Education, or a Fair Chance for the Girls* (Boston: Houghton, Mifflin, 1873). For biographical information on Clarke, see Howard A. Kelly and Walter L. Burrage, *American Medical Biographies* (Baltimore: Norman, Remington, 1920): 225–26; Thomas Francis Harrington, *The Harvard Medical School: A History, Narrative and Documentary* (New York: Lewis, 1905): 868–71. For similar arguments on the higher education of women, see T.S. Clouston, M.D., "Female Education from a Medical Point of View," Part I, PSM 24 (December 1883): 214–28; Clouston, part II, PSM 24 (January 1884): 319–34; A. Hughes Bennett, M.D., "Hygiene in the Higher Education of Women," PSM 16 (February 1880): 519–30.

64. Abby May, "Work Committee," in *Report of the Annual Meeting of the New England Women's Club*, May 31, 1873 (Boston: Rand, Avery, and Company, 1873). History of Women microfilm collection, reel 940, residing at both SSC and SLRI.

65. For an excellent study of the controversy regarding *Sex in Education*, see Mary Roth Walsh, *"Doctors Wanted: No Women Need Apply:" Sexual Barriers in the Medical Profession, 1835–1975"* (New Haven: Yale University Press, 1977): 119–35. See also Crista Deluzio, *Female Adolescence in American Scientific Thought, 1830–1930*, New Studies in American Intellectual and Cultural History series, ed. Howard Brick (Baltimore: Johns Hopkins University Press, 2007), esp. ch. 2 "'Persistence' versus 'Periodicy': From Puberty to Adolescence in the Late Nineteenth-Century Debate over Coeducation," 50–89.

66. M. Carey Thomas, "Present Tendencies in Women's College and University Education," *Educational Review* 25 (1908): 68, quoted in Walsh, *"Doctors Wanted,"* 124, n. 36. For a biography of M. Carey Thomas, see Helen Lefkowitz Horowitz, *The Power and the Passion of M. Carey Thomas* (New York: Alfred A. Knopf, 1994).

67. Edward H. Clarke, *The Building of a Brain* (Boston: J.R. Osgood and Company, 1874), 53.

68. Clarke, *Sex in Education*, 40.

69. Clarke, *Sex in Education*, 93.

70. Clarke, *Sex in Education*, 9.

71. George Fish Comfort and Anna Manning Comfort, *Woman's Education and Woman's Health* (Syracuse: Durston, 1874), 15–16.

72. Mrs. E. [Eliza] B. [Bisbee] Duffey, *No Sex in Education; Or, An Equal Chance for Both Girls and Boys, Being a Review of Dr. E.H. Clarke's 'Sex in Education"* (Philadelphia: J.M. Stoddard, 1874), 117–8.

73. Monograph responses to Clarke included the novel, SOLA (Olive San Louie Anderson), *An American Girl and Her Four Years in a Boy's College* [New York: D. Appleton and Company, 1878]. This was a fictional account of Anderson's experience as one of the first women at the University of Michigan. The narrator had a great time and ended up marrying a classmate. The PSM gave it an unfavorable review because it did not mention what the girl studied or how this affected her health. *Popular Science Monthly* 12 (March 1878), 632. This novel has been reissued, edited by Elizabeth Israels Perry and Jennifer Ann Price (Ann Arbor: University of Michigan Press, 2006). Other responses to Clarke include Julia Ward Howe, ed., *Sex and Education: A Reply to Dr. E. H. Clarke's "Sex in Education"* (Boston: Roberts Brothers, 1874); William B. Green, *Critical Comments upon Certain Passages in the Introductory Portion of Dr. Edward H. Clarke's Book on "Sex in Education"* (Boston: Lee and Shepherd, 1874); Eliza Bisbee Duffey, *No Sex in Education*; Mary Putnam Jacobi, M.D., *The Question of Rest for Women during Menstruation* (New York: G.P. Putnam's Sons, 1886); Anna Brackett, ed., *The Education of American Girls: Considered in a Series of Essays* (New York: G.P. Putnam's Sons, 1874); Antoinette Brown Blackwell, *The Sexes throughout Nature*, ch. 4; and Elizabeth Cady Stanton, "The Coeducation of the Sexes," in *Papers and Letters Presented at the First Woman's Congress of the Association for the Advancement of Woman* (New York: Mrs. William Ballard, 1874), SLRI. For articles in response to *Sex in Education*, see: Sarah D'Arcy, "Sex in Education," *Woman's Journal*, April 25, 1874; M.G.L., "Sex in

Education," *Woman's Journal*, December 27, 1873; Frances D. Gage, "'Sex in Education' Once More," *Woman's Journal*, January 3, 1874; Lydia Fuller, "Matters and Things in St. Louis," *Woman's Journal*, April 4, 1874; See also, "Clarke's Sex in Education," *Nation* 17, November 13, 1873; "The Co-education Question," *Nation* 16, May 22, 1873.

74. Thomas Wentworth Higginson, in Julia Ward Howe, ed., *Sex and Education*, 44, 34. Higginson was one of the most prominent women's rights activists, previously making a name for himself through his writings, abolition activities, and leading an all-black regiment during the Civil War. His other articles on *Sex in Education* include T.W.H. [Thomas Wentworth Higginson], "The Atlantic Monthly on Scientific Education for Women," *Woman's Journal*, May 30, 1874; Thomas Wentworth Higginson, "Physician and Pedagogue," *Woman's Journal*, January 18, 1873; T.W.H., "Just What We Want," *Woman's Journal*, August 16, 1873; T.W.H. "Sex in Education," *Woman's Journal*, November 8, 1873; T.W.H. "Sex in Education—second paper," *Woman's Journal*, November 15, 1873; T.W.H. "Woman's Education and Health Once More," *Woman's Journal*, June 27, 1874; T.W. Higginson, "Sex in Education," *Woman's Signal*, May 13, 1897.

75. Blackwell, *The Sexes throughout Nature*, 166. See also, Antoinette Brown Blackwell, "Sex and Work—No. 8," *Woman's Journal*, June 20, 1874.

76. "Contemporary Literature: Politics, Sociology, Voyages and Travel," *Westminster Review* 104 (1875), 242–3.

77. See, for example, "Testimony from Colleges," in Howe, *Sex and Education*; and Anna C. Brackett, ed., *The Education of American Girls*, which includes essays by leading women, female educators, and women involved with colleges.

78. Elizabeth Cumings, "Education as an Aid to the Health of Women," PSM 17 (October 1880): 823–7.

79. The Clarke debates coincided with the masculinization and professionalization of gynecology and obstetrics, a development that sought to displace midwives as the traditional health care providers for women. For histories of male involvement in gynecological medicine, see Mary Roth Walsh, *"Doctors Wanted: No Women Need Apply"*; and Monica H. Green, *Making Women's Medicine Masculine: The Rise of Male Authority in Pre-Modern Gynecology* (New York: Oxford University Press, 2008).

80. Mrs. E. B. Duffey, *What Women Should Know, A Woman's Book about Women, Containing Practical Information for Wives and Mothers* (Philadelphia: JM Stoddart, 1873; reprint, Sex, Marriage and Society Series, ed. Charles Rosenberg and Carroll Smith-Rosenberg, New York: Arno Press, 1974). Page citations are to the reprint edition.

81. Duffey, *What Women Should Know*, 17–19.

82. Duffey, *What Women Should Know*, 239.

83. Mary Putnam Jacobi, *The Question of Rest for Women during Menstruation* (New York: G.P. Putnam's Sons, 1877), reprinted opposite the title page.

84. Carla Bittel, *Mary Putnam Jacobi and the Politics of Medicine in Nineteenth-Century America*, 126.

85. Jacobi, *Question of Rest*, 227.

86. Bittel, *Mary Putnam Jacobi*, 122.

87. Mary Putnam Jacobi, "Mental Action and Physical Health," 258, in Brackett, ed., *Education of American Girls*, quoted in Bittel, p. 124, fn. 29.

88. Bittel, *Mary Putnam Jacobi*, 127.

89. Bittel, *Mary Putnam Jacobi*, 133–4, 234. Bittel discusses Jacobi's legacy in the context of ongoing discussions of the biology of sex differences and the feminist potential of science in her epilogue, 226–34. For additional analysis of the feminism, science, and female doctors in the nineteenth century, see Regina Morantz-Sanchez, "Feminism, Professionalism, and Germs: The Thought of Mary Putnam Jacobi and Elizabeth Blackwell," *American Quarterly* 34 (1982): 459–78; and Regina Morantz-Sanchez, "Feminist Theory and Historical Practice: Rereading Elizabeth Blackwell," *History and Theory* 31 (1992): 51–69.

90. Several of Gardener's speeches and essays from this time were published in the collection, *Men, Women, Gods,* with an introduction by Robert Ingersoll, 4th ed. revised and enlarged (New York: Truth Seeker Company, 1885), copy residing in the Rare Books Collection, Huntington Library, San Marino, California.

91. Edward H. Clarke, *The Building of a Brain*, 62. For a review of *The Building of a Brain*, see PSM 6 (November 1874): 115–7.

92. Clarke, *The Building of a Brain*, 19.

93. Hammond served from 1861 till August 1864 when he was removed on account of a scandal. He published a pamphlet in his own defense, and the scandal seems to have subsided as no one mentioned it in the 1870s or 1880s. For a report of the scandal, see Samuel Francis, "Biographical Sketches of Distinguished Living New York Physicians," *Medical and Surgical Reporter*, March 2, 1867, p. 165. For a detailed biography of Hammond, see Bonnie Ellen Blustein, *Preserve Your Love for Science: Life of William Hammond, American Neurologist* (New York: Cambridge University Press, 1991); she discusses Gardener's response to Hammond on page 196.

94. Clarke, *The Building of a Brain*, 131–2.

95. William A. Hammond, "Brain-Forcing in Childhood," PSM 30 (April 1887), 731.

96. William A. Hammond, "Woman in Politics," *North American Review* 137 (August 1883), 138–9.

97. Clarke, *The Building of a Brain*, 142.

98. Although Hammond was most prominently associated with the brain size argument, many others advanced similar claims. See also Ely Van de Warker, "Sexual Cerebration," PSM 7 (July 1875): 287–301 and Ely Van de Warker, *Woman's Unfitness for Higher Coeducation* (New York: Grafton Press, 1903); "Relation of Brain-Bulk to Intelligence," PSM 14 (February 1879), 551; H.W.B. "Size of Brain and Size of Body," PSM 16 (April 1880): 827–31; G. Delauney, "Equality and Inequality in Sex," PSM 20 (December 1881): 184–92; J.P.H. Boileau, "Brain-Weight and Brain-Power," PSM 22 (December 1882): 172–4; and Mrs. Z. D. Underhill, "A Premature Discussion," PSM 22 (July 1882): 376–8. For secondary sources on Hammond-Gardener brain debates, see Cynthia Eagle Russett, *Sexual Science*, 38–9; and Kathi Kern, "Gray Matters: Brains, Identities, and Natural Rights," in *The Social and Political Body*, ed. Theodore R. Schatzki and Wolfgang Natter (New York: Guilford Press, 1996): 103–22. For an analysis of how the brain size debates (not related to Hammond and Gardener) impacted working- and lower-class white men, see Rob Boddice, "The Manly Mind? Revisiting the Victorian 'Sex in Brain' Debate," *Gender & History* 23 (August 2011): 321–40. For analysis of brain and gender controversies not necessarily related to Hammond, see also Anne Fausto-Sterling, "A Question of Genius: Are Men Really Smarter Than Women?" in *Myths of Gender:*

*Biological Theories about Women and Men* (New York: Basic Books, 1985): 13–60; and Janet Sayers, *Biological Politics: Feminist and Anti-Feminist Perspectives* (New York: Tavistock Publications, 1982).

99. The earliest feminist reply to Hammond was printed in May 1868. M.C.A. "A Woman's Letter from Washington," *Independent*, May 14, 1868, p. 1. Several women also responded to his essay "Woman in Politics" in the *North American Review*. See, Mrs. L. [Lillie] D. [Devereux] Blake, Nina Morais, Sara A. Underwood, and Clemence Sophia Lozier, "Dr. Hammond's Estimate of Woman," *North American Review* 137 (September 1883): 495–519.

100. Hammond, "Woman in Politics," 139.

101. "Brain size," *Woman's Tribune*, August 1883, p. 3. This article listed the comparative brain weights of various people, including the fact that the heaviest brain on record was that of an illiterate brick layer, to discount Dr. Hammond's argument. See also Antoinette Brown Blackwell, "Comparative Mental Power of the Sexes Physiologically Considered," paper delivered at the 1876 AAW Annual Meeting, Miscellaneous Organizations Collection, box 10, folder 4, SSC; Antoinette Brown Blackwell, "Sex and Work—No. 6," *Woman's Journal*, April 18, 1874; and "Woman's Brain," address of Dr. Kolbenheier before the St. Louis Woman Suffrage Association reprinted in the *Woman's Tribune*, January 24, 1891. Kolbenheier argued that women's brain development had been hindered by cultural obstacles and that their request for more rights was a sign of their evolutionary advance. He favored women's education because it would not only return equilibrium to the sexes but also make for better offspring. Elizabeth Oakes Smith, "Biology and Woman's Rights," an essay read at the Woman Suffrage Convention in Washington, D.C., January 10–12, 1879, reprinted in the *Index*, February 20, 1879. Oakes was responding to an article in PSM (December 1878) about biological arguments against women including brain size. "Edenic Godliness," *Woman's Tribune*, June 11, 1898; "Brains and Sex," *Open Court*, August 18, 1887, 379–80; "Women's Brains Again," *Woman's Tribune*, May 26, 1907. This last article referenced the recent studies of Dr. Roese of Berlin, noting, "Every little while somebody settles the woman question with the assertion that women's brains are smaller than those of men; ergo, they are inferior."

102. Blackwell, "Comparative Mental Power of the Sexes Physiologically Considered," paper delivered at the 1876 AAW Annual Meeting, Miscellaneous Organizations Collection, box 10, folder 4, SSC.

103. Elizabeth Cady Stanton, "The Disabilities and Limitations of Sex" address given at the 1885 National Woman Suffrage Association convention, reprinted in the *Woman's Tribune*, March 1885.

104. The antebellum science of phrenology, for example, was frequently enlisted as proof of the natural inferiority of both women and African Americans. For a related study of early nineteenth century skull collectors and their deterministic views on race, see Ann Fabian, *The Skull Collectors: Race, Science, and America's Unburied Dead* (Chicago: University of Chicago Press, 2010).

105. In addition to being reprinted in *Popular Science Monthly*, Hammond's article received wide coverage. See, for example, "Child or System—Which Shall Live?"

*Christian Union,* April 14, 1887, p. 13; and "The Tyranny of the School," *Philadelphia Medical Times,* April 2, 1887, p. 444.

106. Hammond, "Brain Forcing," 730.

107. Gardener, "Sex and Brain Weight," PSM 31 (June 1887): 266

108. Gardener, "Sex in Brain," 109.

109. Gardener, "Sex in Brain," 104.

110. Gardener, "Sex in Brain," 107.

111. "Dr. Hammond and Woman's Brain," *Woman's Tribune,* January 15, 1889.

112. Hammond, "Men's and Women's Brains," PSM 31 (August 1887): 555.

113. Helen Hamilton Gardener, "Sex and Brain-Weight," 266–8; Hammond replied with "Men's and Women's Brains," PSM 31 (August 1887): 554–8; Gardener responded, "More about Men's and Women's Brains," PSM 31 (September 1887): 698–700; Hammond had the final word with "An Explanation," PSM 31 (October 1887): 846.

114. Hammond, "Men's and Women's Brains," 558.

115. Gardener, "More About," 698.

116. For an analysis of Darwin's use of anthropology, see Rosemary Jann, "Darwin and the Anthropologists: Sexual Selection and Its Discontents," *Victorian Studies* 37, no. 2 (Winter 1994): 287–306.

117. Hammond, "Men's and Women's Brains," 554–8.

118. Kathi Kern also analyzes Gardener's "Sex in Brain" speech, its effect on Stanton, and its larger significance in *Mrs. Stanton's Bible* (Ithaca, NY: Cornell University Press, 2001), 106–9.

119. Gardener, "Sex in Brain," 100.

120. Gardener, "Sex in Brain," 102.

121. Gardener, "Sex in Brain," 101.

122. Gardener, "Sex in Brain," 102.

123. Gardener, "Sex in Brain," 105–6.

124. Gardener, "Sex in Brain," 99.

125. Gardener, "Sex in Brain," 123–4.

126. Elizabeth Cady Stanton, remarks, *Report of The International Council of Women* (Boston: Rufus Darby Printers, 1888), 431. Copy residing at the Center for the History of Medicine, Countway Library, Harvard Medical School.

127. The closeness of their relationship has long been obscured by the apparent disproval of Gardener by Stanton's family. Their close friendship, at least from Gardener's perspective, is indicated by the fact that Stanton asked Gardener to deliver her eulogy and by a letter from Gardener to Burt Wilder, the director of the Wilder Brain Collection at Cornell University, Gardener to Wilder, October 27, 1915, Department of Zoology records, box 17, Cornell University Archives.

128. Stanton's brain bequest form can be found in the Burt Wilder Collection, box 2, folder 4, Cornell University Archives. Lori Ginzberg uncovered it while working on her biography of Stanton; previous historians thought perhaps Gardener had overstated Stanton's desire to donate her brain, which is what Stanton's family believed. Helen Hamilton Gardener to Elizabeth Cady Stanton, July 25, 1887, film, reel 25, *The Papers of Elizabeth Cady Stanton and Susan B. Anthony,* ed. Patricia Holland and Ann D. Gor-

don (Wilmington, DE: Scholarly Resources 1991); quoted in Lori D. Ginzberg, *Elizabeth Cady Stanton: An American Life* (New York: Hill and Wang, 2009), 185, fn. 92. For other treatments of Stanton's contentious brain bequest and Gardener's role in it, see Ellen DuBois, *Harriot Stanton Blatch and the Winning of Woman Suffrage* (New Haven: Yale University Press, 1997), 86; and Kathi Kern, *Mrs. Stanton's Bible*, 108–9, and 250, fn. 58.

129. Helen Hammond Gardener, Eulogy for Elizabeth Cady Stanton, reprinted in the *Woman's Tribune*, November 22, 1902, p. 1.

130. "No, Not Mrs. Stanton's Brain," unidentified clipping, November 1902, Susan B. Anthony Papers, Manuscripts Division, Library of Congress, Washington, D.C.; quoted in Du Bois, *Harriot Stanton Blatch*, 86, fn. 112.

131. Alice Lee, "Data for the Problem of Evolution in Man. A First Study of the Correlation of the Human Skull," *Philosophical Transactions of the Royal Society of London* 69 (1902): 196A; Quoted in Russett, *Sexual Science*, 164–6, fn 17. See also, Rosaleen Love, "'Alice in Eugenics-land': Feminism and Eugenics in the Scientific Careers of Alice Lee and Ethel Elderton," *Annals of Science* 36 (March 1979): 145–58.

132. Franklin Mall, "On Several Anatomical Characters of the Brain Said to Vary according to Race and Sex, with Especial Reference to the Weight of the Frontal Lobe," *American Journal of Anatomy* 9 (1909): 27, 32. Quoted in Russett, *Sexual Science*, 165–6, fn. 19. In *Myths of Gender*, Anne Fausto-Sterling also attributes the demise of the brain weight theory of intelligence to Franklin Mall's 1909 study (38).

133. The transport and procedure for removing Gardener's brain are described in the extensive study of it, James W. Papez, "The Brain of Helen Hamilton Gardener," *American Journal of Physical Anthropology* 11 (October 1927): 29–79. For a history of brain collections, including Cornell's, see Brian Burrell, *Postcards from the Brain Museum: The Improbable Search for Meaning in the Matter of Famous Minds* (New York: Broadway Books, 2004). This book, however, does not mention Gardener.

134. Helen Hamilton Gardener, Last Will and Testament, also reprinted in her funeral booklet, 27–28, Helen Hamilton Gardener Papers (Woman's Rights Collection), SLRI.

135. Papez, "The Brain," 48.

136. Papez, "The Brain," 75.

137. Papez, "The Brain," 63.

138. Papez, "The Brain," 75, 71.

139. Papez, "The Brain," 39. For an analysis of the role of analogy in science—specifically the nineteenth-century analogy that women were like "inferior" races and vice versa—and how this analogy was invoked in brain science and other studies, see Nancy Leys Stepan "Race and Gender: The Role of Analogy in Science," *Isis* 77 (June 1986): 261–77.

140. Papez, "The Brain," 32.

141. Papez, "The Brain," 64.

142. Papez, "The Brain," 64–5.

143. Kathi Kern makes a similar argument in her analysis of Gardener and Stanton's interest in brain donation, see Kern, *Mrs. Stanton's Bible*, 106–9; and also Kern, "Gray Matters," 103–22.

144. Gardener, "Sex in Brain," 107.

145. "Woman's Brain Not Inferior to Men's," *New York Times*, September, 29, 1927, p. 1. See also, "Woman Wills Brain for Research Work," *New York Times*, August 4, 1925, p. 1; and "Says Brain Bequest Has Been Fulfilled," *New York Times*, August 5, 1925, p. 3.

146. "Brain," *Time*, September 14, 1925, p. 12; "Cornell Brain," *Time*, October 10, 1927, p. 28; and "Why You May Wear a Small Hat and Still Have a Big Mind," PSM 107 (December 1925): 15–6, 149. The subtitle of the *Popular Science Monthly* article read "Recent Discoveries about Human Brains and the Minds They Hold—How a Famous Woman, in a Strange Bequest, Sought to Throw New Light on the Old Question: Are Women as Smart as Men?" Interestingly, this article makes no mention of Gardener's earlier battles with Hammond in the pages of *Popular Science Monthly* and, instead, mistakenly notes that she took on Dr. Edward A. Spitzka's ideas about female inferiority.

147. The women's paper *Equal Rights* reprinted a short blurb about Gardener's brain donation from the *Washington Post* in the "Press Comment" section on November 28, 1925, p. 335. Hollingworth and Thompson were leaders in the first generation of female social scientists who established that culture plays a key role in sex differentiation. While their work did not receive much coverage in the women's press, it was written about in the socialist press; see Floyd Dell, "The Nature of Woman," *Masses*, January 1916, p. 16. For an excellent study of their efforts to refute the theory of greater male variability, see Rosalind Rosenberg's *Beyond Separate Spheres: Intellectual Roots of Modern Feminism* (New Haven: Yale University Press, 1982). See also, Stephanie Shields, "The Variability Hypothesis: The History of a Biological Model of Sex Differences in Intelligence," *Signs* 7 (Summer 1982): 769–97.

148. For example, it seems as if there are new books about the differences between male and female brains published each year, drawing on the latest scientific techniques and often describing differences much like those that Gardener hoped to discredit. For two that have received a tremendous amount of press coverage and become national best sellers, see Louann Brezendine, *The Female Brain* (New York: Broadway Books, 2006); and Louann Brezendine, *The Male Brain* (New York: Broadway Books, 2010).

149. Helen Hamilton Gardener, Last Will and Testament, SLRI.

150. Wendy Kline analyzes the challenges and opportunities that activists in the 1960s and 1970s women's health movement faced when attempting to base their arguments on the idea that women could be unified around their bodies and their reproductive health in *Bodies of Knowledge: Sexuality, Reproduction, and Women's Health in the Second Wave* (Chicago: University of Chicago Press, 2010).

CHAPTER 3

1. Judith Allen charts this fascinating period of Gilman's life in *The Feminism of Charlotte Perkins Gilman: Sexualities, Histories, Progressivism* (Chicago: University of Chicago Press, 2009), 43–63. For additional biographies of Gilman, see, Helen Lefkowitz Horowitz, *Wild Unrest: Charlotte Perkins Gilman and the Making of "The Yellow Wall-Paper"* (New York: Oxford University Press, 2012); Cynthia J. Davis, *Charlotte Perkins Gilman: A Biography* (Stanford: Stanford University Press, 2010); Mary A. Hill,

*Charlotte Perkins Gilman: The Making of a Radical Feminist, 1860–1896* (Philadelphia: Temple University Press, 1979); and Gary Scharnhorst, *Charlotte Perkins Gilman*, Twayne's United States Authors Series (Boston: Twayne, 1985). Scharnhorst has also compiled a comprehensive bibliography of Gilman's writing: *Charlotte Perkins Gilman: A Bibliography* (Metuchen: Scarecrow, 1985). See also, Jill Rudd and Val Gough, eds., *Charlotte Perkins Gilman: Optimist Reformer* (Iowa City: University of Iowa Press, 1999).

2. Charlotte Perkins Gilman, *The Living of Charlotte Perkins Gilman: An Autobiography* (1935), introduction by Ann J. Lane (Madison: University of Wisconsin Press, 1990), 143, 163.

3. Carl Degler, introduction to Charlotte Perkins Gilman, *Women and Economics: A Study of the Economic Relation between Men and Women as a Factor in Social Evolution* (Boston: Small, Maynard and Company, 1899), Harper Torchbook edition (New York: Harper and Row, 1966), xxvi.

4. For example, in his discussion of the mental differences between men and women on page 629, Darwin refers to women's maternal instincts as resulting in "her greater tenderness and less selfishness." He also lists "maternal instinct" as one of the abiding common denominators connecting humans with the lower animals (131–7).

5. In her study of American attitudes toward female domesticity, Glenna Matthews argues that Darwinism, together with industrialization, destroyed the earlier nineteenth-century appreciation for female household labor. She claims that Darwin's suggestion that females' main contribution to evolution was reproductive encouraged other intellectuals to consider household work insignificant. Glenna Mathews *"Just a Housewife": The Rise and Fall of Domesticity in America* (New York: Oxford University Press, 1987), see esp. ch. 5: "Darwinism and Domesticity: The Impact of Evolutionary Theory on the Status of the Home," 116–44.

6. See, for example, Gail Bederman, "'Not to Sex—but to Race!' Charlotte Perkins Gilman, Civilized Anglo-Saxon Womanhood, and the Return of the Primitive Rapist," in *Manliness and Civilization: A Cultural History of Gender and Race in the United States, 1880–1917*, Women in Culture and Society Series, ed. Catharine R. Stimpson (Chicago: University of Chicago Press, 1995): 121–69

7. Ideas for cooperative housework—which Gilman opposed in preference of paid housework done by professionals—circulated among late nineteenth-century feminists and reformers. For other reformers who suggested cooperative housework as part of larger feminist demands, see Abby Morton Diaz, *The Schoolmaster's Trunk* (1874) and Marie Howland, *Papa's Own Girl* (1874). Both novels are mentioned as examples of demands for cooperative housework in Sylvia Strauss, "Gender, Class, and Race in Utopia," in *Looking Backward: 1988–1888: Essays on Edward Bellamy*, ed. Daphne Patai (Amherst: University of Massachusetts Press, 1988), 68–90. Cooperative housework was also a feature of the utopia imaged by Edward Bellamy in *Looking Backward* (1888).

8. The most popular domestic advice book of the nineteenth century was written by Gilman's aunt, Catharine E. Beecher, *A Treatise on Domestic Economy for the Use of Young Ladies at Home, and at School* (Boston: Marsh, Capen, Lyon, and Webb, 1841) For more on Beecher, see Kathryn Kish Sklar, *Catharine Beecher: A Study in American Domesticity* (New Haven: Yale University Press, 1973).

9. Samuel K. Jennings, *The Married Lady's Companion, or Poor Man's Friend*, 2d ed. (New York: J.C. Totten, 1808, reprint, Medicine and Society in America Series, ed. Charles Rosenberg, New York: Arno Press, 1972), 76–77 (page citations are to the reprint edition).

10. Jennings, *The Married Lady's Companion*, 8.

11. E. [Elizabeth] C. [Cady] S.[Stanton], "Our Young Girls," *Revolution*, January 29, 1868, p. 57.

12. E.T. Grover, "Woman and the Curse," *Woman's Tribune*, January 1887, p. 2.

13. Elizabeth Cady Stanton, "The Woman's Bible, chapter III," *Woman's Tribune*, April 6, 1895, p. 1.

14. In the second half of the nineteenth century, women increasingly turned to experts, generally male physicians, for advice about pregnancy and child rearing. Previously considered something women intuitively knew, motherhood became something for which women needed education and training. For a broader discussion of expert intervention in motherhood, see Rima D. Apple, *Perfect Motherhood: Science and Childrearing in America* (New Brunswick: Rutgers University Press, 2006). See also Barbara Ehrenreich and Deirdre English, *For Her Own Good: Two Centuries of the Experts' Advice to Women*, second Anchor Books edition (New York: Anchor Books, 2005).

15. Monfort Allen, M.D., and Amelia McGregor, M.D., *The Woman Beautiful, or Maidenhood, Marriage, and Maternity* (Philadelphia: National Publishing, 1901), 44–45.

16. Mary Wood-Allen, M.D., *Almost a Woman*, Teaching Truth Series (Cooperstown: Arthur H. Crist, 1911), 59. Copy residing at the Center for the History of Medicine, Countway Library, Harvard Medical School (CHM).

17. Marion Harland, *Eve's Daughters or Common Sense for Maid, Wife and Mother* (New York: John R. Anderson, 1882), 422, 437, 441. *Popular Science Monthly* favorably reviewed *Eve's Daughters* in volume 23 (September 1883), 711.

18. E. [liza] B. [isbee] Duffey, *What Women Should Know, A Woman's Book about Women, Containing Practical Information for Wives and Mothers* (Philadelphia: JM Stoddart, 1873; reprint, Sex, Marriage and Society Series, ed. Charles Rosenberg and Carroll Smith-Rosenberg, New York: Arno Press, 1974), 43. Page citations are to the reprint edition.

19. Duffey, *What Women Should Know*, 252. See also Pye Henry Chavasse, M.D., *Wife and Mother or Information for Every Woman* (Philadelphia: HJ Smith and Company, 1888); and Burt Wilder, *What Young People Should Know: The Reproductive Function in Man and the Lower Animals* (Boston: Estes and Lauriat, 1876).

20. In Western societies, there is a long history of ideas about women's maternity mirroring larger discussions of natural and political rights. For analysis of these relationships, see, for example, Thomas Laqueur, "Orgasm, Generation and the Politics of Reproductive Biology," in Thomas Laqueur and Catherine Gallagher, ed., *The Making of the Modern Body: Sexuality and Society in the Nineteenth Century* (Berkeley: University of California Press, 1987): 1–41; and Londa Schiebinger, "Why Mammals Are Called Mammals," in Londa Schiebinger, *Nature's Body: Gender in the Making of Modern Science* (New Brunswick, NJ: Rutgers University Press, 2004): 40–74.

21. Elizabeth Cazden, *Antoinette Brown Blackwell: A Biography* (Old Westbury, NY: Feminist Press, 1983), 56–57, 267.

22. Antoinette Brown Blackwell, "Marriage and Work," 1875 Congress of the Association for the Advancement of Women, held in Chicago, p. 28. Miscellaneous Organizations Collection, box 10, folder 3, Sophia Smith Collection, Smith College, Northampton, Massachusetts (SSC).

23. Blackwell with Mrs. Claude U. Gilson, unpublished autobiography, ch. 11, "Feminism and Babies," 223. Blackwell Family Papers, Schlesinger Library, Radcliffe Institute, Harvard University, Cambridge, Massachusetts (SLRI).

24. Blackwell with Gilson, unpublished autobiography, 232.

25. Blackwell with Gilson, unpublished autobiography, 233.

26. Blackwell belonged to the Association for the Advancement of Women (AAW) and frequently appeared in its programs. For example, she delivered a presentation on "How to Combine Intellectual Culture with Household Management and Family Duty," at the second annual AAW congress in Chicago in 1874. She also delivered a paper entitled "Where Is the Work of Women Equal, Where Superior, Where Inferior to That of Men," at the AAW meeting in 1888, Miscellaneous Organizations Collection, box 10, folder 10, SSC. The fifteenth annual convention booklet lists all the presentations from the previous AAW congresses, Fifteenth Annual Convention of the Association for the Advancement of Women, October 26, 1887, Sorosis records, box 5, folder 9, SSC. Blackwell also delivered papers at the Association for the Advancement of Science and frequently published in the *Woman's Journal*. See, for example, Blackwell, "The Comparative Longevity of the Sexes," read before the American Association for the Advancement of Science, Philadelphia, August 1884, copy residing at the CHM.

27. Sarah Blaffer Hrdy, *The Woman That Never Evolved* (Cambridge: Harvard University Press, 1981), ix and 13.

28. Cazden discusses Blackwell's belief in human distinctiveness on p. 147.

29. Blackwell, *Sexes*, 37.

30. Blackwell, *Sexes*, 73.

31. Blackwell, *Sexes*, 20.

32. Helen Hamilton Gardener, "Woman as an Annex" reprinted in Gardener, *Facts and Fictions of Life* (Boston: Arena Publishing, 1895), 132.

33. Helen Hamilton Gardener, *Plain Talk: A Pamphlet on the Population Question and the Moral Responsibility of Woman in Maternity* (Chicago: G.E. Wilson, 19—), 11–12. History of Women Collection.

34. Gardener, "Woman as an Annex," 129.

35. Gardener, "Woman as an Annex," 129.

36. Blackwell, *Sexes*, 55–8.

37. Blackwell, *Sexes*, 60–61.

38. Although, as several historians have noted, by the 1890s rates of breast-feeding among white, middle- and upper-class women fell precipitously. Adrienne Berney suggests that this decline was related to cultural changes, including the fear of close association with animals—a fear made all too real by Darwinian evolution—and by racial ideologies that linked wet nursing with African American and immigrant women. See Adrienne W. Berney, "Reforming the Maternal Breast: Infant Feeding in American Culture, 1870–1920" (Ph.D. diss., University of Delaware, 1998). See also Rima D. Apple, *Mothers and Medicine: A Social History of Infant Feeding, 1890–1950*, Wisconsin

Publications in the History of Science and Medicine, no. 7 (Madison: University of Wisconsin Press, 1987); Bernice Hausman, *Mother's Milk: Breastfeeding Controversies in American Culture* (New York: Routledge, 2003); and Jacqueline H. Wolf, *Don't Kill Your Baby: Public Health and the Decline of Breastfeeding in the Nineteenth and Twentieth Centuries* (Columbus: Ohio State University Press, 2001).

39. Blackwell, *Sexes*, 30–31.

40. Blackwell, *Sexes*, 30–31.

41. Reverend Wilder Smith, "The Descent of Man from the Apes," *Minnesotan Herald*, June 10, 1871, p. 4.

42. "The Sexes throughout Nature," *Popular Science Monthly* 7 (July 1875), 371. For other reviews of Blackwell's science books, see also "Our Book Table," *Inter Ocean* (Chicago) May 1, 1875; *Harper's New Monthly Magazine* 39 (July 1869), 290; "Sexes throughout Nature, *Evening Bulletin* (San Francisco), June 12, 1875, p. 1.

43. Blackwell, *Sexes*, 33.

44. Blackwell, *Sexes*, 137.

45. Blackwell, "Work in Relation to the Home," *Woman's Journal*, May 2, 1874.

46. Blackwell, "Work in Relation to the Home."

47. Blackwell, "Where Is the Work of Women Equal," 84.

48. Blackwell, *Sexes throughout Nature*, 112.

49. Blackwell, "Where Is the Work of Women Equal," 85.

50. Cazden critiques Blackwell for catering to the needs and desires of middle- and upper-class women who could afford to hire (female) servants, as Blackwell herself often did, and for failing to consider women whose work would take them out of the home on a daily basis (165, 168, 185). But my reading of Blackwell's vast writing on the topic suggests that, while she may have hired help in her own home, she prioritized greater male involvement.

51. Blackwell, *Sexes*, 112–3.

52. Blackwell, *Sexes*, 113–6.

53. Blackwell with Gilson, unpublished autobiography, 211.

54. Cazden, *Blackwell*, 165–9, 185.

55. Blackwell, *Sexes*, 167.

56. Blackwell with Gilson, unpublished autobiography, 191.

57. Blackwell, *Sexes*, 235.

58. Antoinette Brown Blackwell, "Work in Relation to the Home, II," *Woman's Journal*, May 9, 1874. She also delivered a speech about the division of labor and sharing household responsibilities at the first AAW Congress held in October 1873 in New York City, *Papers and Letters Presented at the First Woman's Congress* (New York: Mrs. WM. Ballard Printer, 1874). Copy residing at the SLRI.

59. "Sexes throughout Nature," [San Francisco] *Evening Bulletin*, June 12, 1875, pp. 1–2.

60. Blackwell, *Sexes*, 135.

61. In addition to Blackwell and Lucy Stone (the wife of Henry Blackwell), were the Blackwell sisters Emily and Elizabeth, both doctors.

62. In 1870, seven out of ten women who worked were servants, according to Nancy Woloch, *Women and the American Experience*, 5th ed. (New York: McGraw Hill, 2011), 216.

63. Woloch, *Women*, 279.

64. Woloch, *Women*, 216.

65. Rebecca Edwards, *New Spirits: Americans in the Gilded Age, 1865–1905* (New York: Oxford University Press, 2006), 135.

66. Woloch, *Women*, 280–81.

67. Woloch, *Women*, 118. In contrast, African American birthrates did not begin to drop until the 1880s and immigrant birthrates remained steady.

68. Grant Allen, "Plain Words on the Woman Question," *Popular Science Monthly* 36 (December 1889): 170–81.

69. "Biology versus Nature," *Woman's Standard* 4.5 (January 1890): 4–5.

70. Mary Jordan Finley, "Evolution's Solution to the Problem," *Lucifer the Light-Bearer*, August 14, 1896. Throughout 1890, *Lucifer* also published a multipart series on "The Woman Question" by Penelope that focused on refuting the claims of Grant Allen.

71. Charles Howard Fitch, "Woman's Place in Nature," *Woman's Tribune*, January 4, 1890, p. 58. The *Woman's Tribune* ran numerous responses to Allen during 1889–1892.

72. Louise Fiske-Bryson, M.D., "Woman and Nature," *New York Medical Journal*, December 3, 1887, p. 627.

73. For an extended analysis of Ross, see Laura Lovett, *Conceiving the Future: Pronatalism, Reproduction, and the Family in the United States, 1890–1938* (Chapel Hill: University of North Carolina Press, 2007), ch. 4, "The Political Economy of Sex: Edward A. Ross and Race Suicide," 77–108.

74. Gail Bederman, *Manliness and Civilization*, 200–206.

75. For a history of the WCTU, see Ian Tyrrell, *Woman's World/Woman's Empire: The Woman's Christian Temperance Union in International Perspective, 1880–1930* (Chapel Hill: University of North Carolina Press, 1991).

76. Christine Woyshner, "Race, Gender, and the Early PTA: Civic Engagement and Public Education, 1897–1924," *Teachers College Record* 105 (April 2003): 520–44. See also chapter 1 in Ann Hulbert, *Raising America: Experts, Parents, and a Century of Advice about Children* (New York: Alfred A. Knopf, 2003).

77. Charlotte Perkins Gilman, *Women and Economics: A Study of the Economic Relation between Men and Women as a Factor in Social Evolution*, edited and with an introduction by Carl Degler (Boston: Small, Maynard, and Company 1898, reprint New York: Harper Torchbook, 1966;), 5, 37 (page citations are to the reprint edition). For further analyses of the role of evolution in Gilman's thought, see Carl Degler's introduction to the 1966 edition of *Women and Economics*, as well as Carl Degler, "Charlotte Perkins Gilman on the Theory and Practice of Feminism," *American Quarterly* 8 (Spring 1956): 21–39; Gail Bederman, "Not to Sex—but to Race!"; Maureen L. Egan, "Evolutionary Theory in the Social Philosophy of Charlotte Perkins Gilman," *Hypatia* 4 (Spring 1989): 102–19; Bernice L. Hausman, "Sex before Gender: Charlotte Perkins Gilman and the Evolutionary Paradigm of Utopia," *Feminist Studies* 24 (Fall 1998): 489–510; and Alys Eve Weinbaum, "Writing Feminist Genealogy: Charlotte Perkins Gilman and the Reproduction of Racial Nationalism," ch. 2 in *Wayward Reproductions: Genealogies of Race and Nation in Transatlantic Modern Thought* (Durham:

Duke University Press, 2004). For an analysis of Gilman's thought in relation to Spencer's, see Lois N. Magner, "Darwinism and the Woman Question: The Evolving Views of Charlotte Perkins Gilman," in *Critical Essays on Charlotte Perkins Gilman*, ed. Joanne Karpinski, Critical Essays on American Literature Series, ed. James Nagel (New York: G.K. Hall, 1992): 115–28. For an additional analysis of Gilman's formulation of the "sexuo-economic" relationship, see Cynthia J. Davis, "Love and Economics: Charlotte Perkins Gilman on 'The Woman Question,'" *ATQ* 19 (December 2005): 243–58.

78. As recounted in, Degler, introduction to *Women and Economics*, xiii, xix.

79. Ella Ormsby, "A Woman's Book," review of *Women and Economics*, *Labor Advocate* (Birmingham, AL), October 14, 1899.

80. Gilman, *Women and Economics*, 1.

81. Allen, *Gilman*, 73.

82. Gilman, *The Living of Charlotte Perkins Gilman*, 35.

83. Gilman, *The Living of Charlotte Perkins Gilman*, 8–9.

84. Bederman, *Manliness*, 145–6.

85. Bederman, *Manliness*, 122.

86. Allen, *Feminism of*, 140, 347. Allen discusses Gilman and the race question, thoroughly reviewing all the relevant scholarship, on pages 338–47.

87. Lester Frank Ward, "Our Better Halves," *Forum* 6 (November 1888): 266–75; and Lester Frank Ward, ch. 14, *Pure Sociology, A Treatise on the Origin and Spontaneous Development of Society* (New York: MacMillan Company, 1903). For more information on Ward's gynaecocentric theory, see Clifford Scott, "A Naturalistic Rationale for Women's Reform: Lester Frank Ward on the Evolution of Sexual Relations," *Historian* 33 (November 1970): 54–67. See also Clifford Scott, *Lester Frank Ward* (Boston: Twayne, 1976). For more on Ward's thinking about gender, see Barbara Finlay, "Lester Frank Ward as a Sociologist of Gender: A New Look at His Sociological Work," *Gender and Society* 13 (April 1999): 251–65. Finlay argues that Ward's work was foundational in the field of sociology but that after his death he was widely forgotten by the field, largely because of his support of feminist and progressive causes.

88. Edwards, *New Spirits*, 19.

89. For a discussion of Ward as a reform Darwinist, see Beryl Satter, *Each Mind a Kingdom: American Women, Sexual Purity, and the New Thought Movement, 1875–1920* (Los Angeles: University of California Press, 1999), 36–7.

90. Clifford Scott, "A Naturalistic Rationale," 64–5.

91. Ward, "Our Better Halves," 269.

92. Ward, "Our Better Halves," 275.

93. Finlay, "Lester Frank Ward," 258–9.

94. Finlay, "Lester Frank Ward," 252.

95. Charlotte Perkins Gilman, dedication page, *The Man-Made World, or Our Androcentric Culture* (New York: Charlton Company, 1911).

96. Gilman,*The Man-Made World*, 13, 25.

97. Gilman, *The Living of Charlotte Perkins Gilman*, 42.

98. Gilman, "Comment and Review," *Forerunner* 12 (October 1912): 26.

99. Gilman, *Women and Economics*, xxxvi–xxxvii.

100. Gilman, *Women and Economics*, 178.

101. Gilman, *Women and Economics*, 189–90.

102. Gilman, *Women and Economics*, 194.

103. Gilman, *Women and Economics*, 240.

104. For an extended discussion of Gilman's plans, as well as their historic antecedents, see Dolores Hayden, *The Grand Domestic Revolution* (Boston: MIT Press, 1981), ch. 9, 182–205.

105. New England Women's Club, Discussion Committee, 1898–1899. New England Women's Club Collection, Discussion Committee, folder 72, SLRI.

106. Review of *Concerning Children* by Charlotte Perkins Gilman, *Clubwoman*, February 1901.

107. Elizabeth Sloan Chesser, *Woman, Marriage, and Motherhood*, with an introduction by Mrs. Frederic Schoff, president of the National Congress of Mothers and Parent-Teacher Association (New York: Funk and Wagnalls, 1913), 9.

108. Chesser, *Woman*, 262–70.

109. For scholarly discussions of the Gilman-Key debates, see Allen, *Feminism of*, 174–84; Ann Taylor Allen, "Feminism, Social Science, and the Meanings of Modernity: The Debate on the Origin of the Family in Europe and the United States, 1860–1914," *American Historical Review* 104 (October 1999): 1085–109; and Nancy Cott, *The Grounding of Modern Feminism* (New Haven: Yale University Press, 1987), 42, 45, 49, 20. According to Mari Jo Buhle's study of women socialists, Katharine Anthony was Ellen Key's promoter in the United States. Buhle, *Women and American Socialism, 1870–1920* (Urbana: University of Illinois Press, 1983), 40; she is described further in Allen, *Feminism of*, 171–80.

110. Allen, *Feminism of*, 174.

111. See Charlotte Perkins Gilman, "On Ellen Key and the Woman Movement," *Forerunner* 4 (February 1913): 35–38. She pointed out that the two shared much common ground, with the exception of whether or not women should devote themselves entirely to motherhood.

112. Ellen Key, *The Century of the Child* (New York: G.P. Putnam and Sons, 1909), 8.

113. Key, *The Century*, 12.

114. Key, *The Century*, 46.

115. Key, *The Century*, 58.

116. Charlotte Perkins Gilman, "The New Motherhood," *Forerunner* 1 (December 1910): 17–18. See also Gilman's review of Key's *Love and Marriage*, Comment and Review, *Forerunner* 2 (October 1911): 280; and her review of *The Woman Movement*, "On Ellen Key and the Woman Movement," *Forerunner* 4 (February 1913): 35–38.

117. Gilman, "On Ellen Key," 36.

118. Charlotte Perkins Gilman, "This 'Superiority,'" *Forerunner* 2 (April 1911): 126.


CHAPTER 4

1. Her two surviving children were Helen Burt Gamble and William Burt Gamble. Helen married Dr. William C. Martin of Detroit; William graduated from the Massa-

chusetts Institute of Technology and, later, led the New York Public Library's science and technology department. "Eliza Burt Gamble," *The National Cyclopaedia of American Biography: Being the History of the United States, Volume 18* (New York: James T. White & Company, 1922), 220–1. While many other nineteenth-century feminists combined their maiden and married last names (for example, Antoinette Brown Blackwell, Elizabeth Cady Stanton, Mary Putnam Jacobi), I can think of no other examples of women passing on their maiden names to their children in this way. Thanks to Katherine Ana Ericksen for leading me to this image of Gamble. Katherine wrote her senior honors thesis on Gamble under the supervision of Professor Sarah Richardson, who put us in touch: Katherine Ana Ericksen, "Eliza Burt Gamble and the Proto-Feminist Engagements with Evolutionary Theory" (A.B. thesis, Honors in History of Science, Harvard University, 2011).

2. Elizabeth Cady Stanton, Susan B. Anthony, and Matilda Joslyn Gage, *History of Woman Suffrage*, vol. 3 (Rochester, NY: Charles Mann Printing, 1886), 516.

3. Eliza Burt Gamble, "Mr. Parkman and the Woman Question" (East Saginaw, MI: Courier Co., Printers and Binders, 1880), History of Women Collection, available at the Schlesinger Library, Radcliffe Institute, Harvard University (SLRI).

4. Eliza Burt Gamble, preface to the first edition of *The Evolution of Woman* (1894), later republished as *The Sexes in Science and History: An Inquiry into the Dogma of Woman's Inferiority to Man* (New York: G.P. Putnam's Sons, 1916), v. Page citations are to *The Sexes*. A handful of scholars have written about Gamble's feminist interpretation of evolution. See, for example, Sally Gregory Kohlstedt and Mark R. Jorgensen, "'The Irrepressible Woman Question': Women's Responses to Evolutionary Ideology," in *Disseminating Darwinism: The Role of Place, Race, Religion, and Gender*, ed. Ronald L. Numbers and John Stenhouse (New York: Cambridge University Press, 1999): 267–93. The most thorough analysis is Rosemary Jann, "Revising the Descent of Woman: Eliza Burt Gamble," in *Natural Eloquence: Women Reinscribe Science*, ed. Barbara T. Gates and Ann B. Shteir, Science and Literature Series, ed. George Levine (Madison: University of Wisconsin Press, 1997): 147–63. Gamble is also discussed in Penelope Deutscher, "The Descent of Man and the Evolution of Woman," *Hypatia* 19 (Spring 2004): 35–55; and Griet Vandermassen, *Who's Afraid of Charles Darwin?: Debating Feminism and Evolutionary Theory* (Lanham: Roman and Littlefield Publishers, 2005).

5. *History of Woman Suffrage*, vol.3, 659

6. Gamble, preface to *The Evolution of Woman* reprinted in *The Sexes*, v.

7. Gamble, *The Sexes in Science and History*, vi.

8. As George Levine, Elizabeth Grosz, and other scholars have pointed out more recently, female choice is a potentially radical concept. "On the strength of it," Levine argues, "one might make a case for the sexist Darwin as a kind of ideological hero in spite of himself. Certainly he believed in the intellectual inferiority of women. But female choice was about as revolutionary a concept as natural selection, and only recently has resistance to it diminished." George Levine, *Darwin Loves You: Natural Selection and the Re-enchantment of the World* (Princeton: Princeton University Press, 2006), 201. For more on alternative readings of sexual selection theory in British literature, see Gillian Beer, ch. 7, "Descent and Sexual Selection: Women in Narrative," in *Darwin's*

*Plots: Evolutionary Narrative in Darwin, George Eliot, and Nineteenth-Century Fiction*, 2d ed. (New York: Cambridge University Press, 2000): 198–219.

9. For a history of the scientific reception of and engagement with female choice, see Erika Lorraine Milam, *Looking for a Few Good Males: Female Choice in Evolutionary Biology* (Baltimore: Johns Hopkins University Press, 2010).

10. Darwin describes male "eagerness" and female "coyness" in *The Descent of Man, and Selection in Relation to Sex*, 2d ed. (1879), with an introduction by Adrian Desmond and James Moore (New York: Penguin Classics, 2004), 256–7.

11. Darwin, *The Descent of Man*, 585.

12. Darwin, *The Descent of Man*, 665–6.

13. The year after publishing *The Descent of Man*, Darwin published *The Expression of the Emotions in Man and Animals* (1872) to further elaborate his ideas on human-animal kinship and, especially, to make clear that he really did mean that animals experience every sort of emotion that humans do.

14. Alfred Russel Wallace, *Darwinism: An Exposition of the Theory of Natural Selection with Some of Its Applications* (New York: MacMillan, 1889). Helena Cronin argues that Wallace and Darwin's debates regarding sexual selection shaped the theory's reception from the publication of *The Descent of Man* till today; see Helena Cronin, *The Ant and the Peacock: Altruism and Sexual Selection from Darwin to Today* (Cambridge: Cambridge University Press, 1991). The most comprehensive analysis of the Wallace-Darwin debates on sexual selection can be found in Milam, *Looking for a Few Good Males*. Milam rejects the "eclipse" narrative of female choice, instead arguing that this narrative became a popularly accepted truth in the 1970s and 1980s because of internecine struggles within biology. For other histories of sexual selection theory and its reception, see Peter J. Vorzimmer, *Charles Darwin: The Years of Controversy, The Origin of Species and Its Critics, 1859–1882* (Philadelphia: Temple University Press, 1970), 191–7; Gertrude Himmelfarb, *Darwin and the Darwinian Revolution* (Garden City: Doubleday, 1962); Michael Ghiselin, *The Triumph of the Darwinian Method* (Berkeley: University of California Press, 1969); Mary Margaret Bartley, "A Century of Debate: The History of Sexual Selection Theory (1871–1971)" (Ph.D. diss., Cornell University, 1994); Bernard Campbell, ed., *Sexual Selection and the Descent of Man 1871–1971* (Chicago: Aldine Publishing Company, 1972), an edition published to commemorate the one hundredth anniversary of the *Descent*; Simon J. Frankel, "The Eclipse of Sexual Selection Theory," in *Sexual Knowledge, Sexual Science: The History of Attitudes to Sexuality*, ed. Roy Porter and Mikuláš Teich (Cambridge: Cambridge University Press, 1994): 158–83; and Kay Harel, "When Darwin Flopped: The Rejection of Sexual Selection," *Sexuality and Culture* 5, no. 4 (Fall 2001): 29–42. For a primary account of the controversies regarding sexual selection, see George Romanes, *Darwin, and after Darwin*, 3d ed. (Chicago: Open Court Publishing Company, 1901), esp. ch. 10.

15. Wallace, *Darwinism*, 286.

16. Milam, *Looking for a Few Good Males*, ch. 1.

17. Charles Darwin, preliminary notice, in W. [William] T. [Thomson] Van Dyck, "On the Modification of a Race of Syrian Street-Dogs by Means of Sexual Selection," *Proceedings of the Zoological Society of London* no. 25 (April 4, 1882): 367–69. Available

online via John van Wyhe, ed. 2002. The Complete Work of Charles Darwin Online (http://darwin-online.org.uk/).

18. Wallace, *Darwinism*, 392.

19. St. George Mivart, "Review of *The Descent of Man*," *Quarterly Review*, 1871, 59.

20. More recent criticism of sexual selection theory has condemned it for conforming to the most egregious examples of Victorian sexism and naturalizing ideas about gender difference and heteronormativity. See, for example, Joan Roughgarden, *The Genial Gene: Deconstructing Darwinian Selfishness* (Los Angeles: University of California Press, 2009); and Joan Roughgarden, *Evolution's Rainbow: Diversity, Gender, and Sexuality in Nature and People, with a New Preface* (Los Angeles: University of California Press, 2009).

21. Lester Frank Ward, "The Past and Future of the Sexes," *Independent*, March 8, 1906, 544–5. Social scientists Thorstein Veblen and W.I. Thomas also critiqued female fashions from an evolutionary perspective, though Veblen relied on an economic as opposed to biological model of evolution. See Thorstein Veblen, *The Theory of the Leisure Class* (1899), reprinted with an introduction by Robert Lekachman (New York: Penguin Books, 1967); and W.I. Thomas, "The Psychology of Women's Dress," *American Magazine* (November 1908), 66. For an analysis of the link between Darwinism and demands for more simple dress, see Valerie Steele, *Fashion and Eroticism: Ideals of Feminine Beauty from the Victorian Era to the Jazz Age* (New York: Oxford University Press, 1985), 148–9. Charlotte Perkins Gilman also extensively critiqued female fashions and lobbied for more practical fashions, including pockets for women, in her journal the *Forerunner*.

22. For more on women's animal-inspired fashions, see Susan David Bernstein, "Designs after Nature: Evolutionary Fashions, Animals, and Gender," ch. 4 in *Victorian Animal Dreams: Representations of Animals in Victorian Literature and Culture* (Burlington: Ashgate Publishing Company, 2007): 65–79; and Jennifer Price, "When Women Were Women, Men Were Men, and Birds Were Hats," ch. 2 in *Flight Maps: Adventures with Nature in Modern America* (New York: Basic Books, 1999): 57–110.

23. Gamble, *The Sexes*, 32.

24. Gamble, *The Sexes*, 38.

25. Gamble, *The Sexes*, 32–33.

26. Gamble, *The Sexes*, 36–37.

27. Gamble, *The Evolution of Woman*, 69–70.

28. Cynthia Eller, *Gentlemen and Amazons: The Myth of Matriarchal Prehistory, 1861–1900* (Berkeley: University of California Press, 2011), 6, 105, 112.

29. Gamble, *Sexes*, 61.

30. In her biography of Margaret Sanger, Ellen Chesler notes that the maternal death rate in the United States used by turn-of-the-century reformers was seven deaths per one thousand live births, one of the highest in the industrialized world. Chesler, *Woman of Valor: Margaret Sanger and the Birth Control Movement in America* (New York: Simon and Schuster Paperback, 2007, 1st ed. 1992), 65. For a comprehensive study of maternal death rates, see Irvine Loudon, *Death in Childbirth: An International*

*Study of Maternal Care and Maternal Mortality 1800–1950* (New York: Oxford University Press, 1993).

31. Eliza Burt Gamble, "Race Suicide in France," in *The International Socialist Review IX (July 1908-June 1909)* (Chicago: Charles H. Kerr & Company, 1909), 513.

32. Gamble, "Race Suicide in France," 513.

33. Eliza Burt Gamble, "Fears the Effect of Pension Idea," Letter to the Editor, *Detroit Free Press*, October 21, 1911, p. 23.

34. Eliza Burt Gamble, "Says Free Women Could Stop Vice," Letter to the Editor, *Detroit Free Press*, October 1, 1913, p. 18.

35. Gamble, *Sexes in Science and History*, 82.

36. For a comprehensive study of nineteenth- and twentieth-century critiques of marriage and women's lack of autonomy within it, see Linda Gordon, *The Moral Property of Women: A History of Birth Control Politics in America* (Chicago: University of Illinois Press, 1974).

37. "Books of the Week," *New York Herald-Tribune*, March 3, 1894, p. 8. See also, "Literary Notes," *San Francisco Chronicle* , February 11, 1894, p. 9.

38. "The Condition of Women," *The Literary World: A Fortnightly Review of Current Literature*, April 21, 1894, p. 119. This review concluded, "She has been an industrious student in both fields [science and anthropology], her matter is well arranged and her positions are clearly put; but her whole volume is a striking example of a kind of reasoning not monopolized by either sex—the imposition of preconceived conclusions upon long-suffering facts." This review compared Gamble's *The Evolution of Woman* with *The Rights of Woman: A Comparative Study in History and Legislation* by M. Ostrogorski, a book arguing against the idea that women could claim natural or political rights. The review praised Ostrogorski for "confin[ing] himself to a statement of undisputed facts and obvious tendencies, and his book has throughout a very judicial quality," in contrast to what the reviewer took to be the biased approach of Gamble. The *Chicago Daily Tribune* voiced a similar concern, noting, "The work is, therefore, largely an argument against the logical conclusions of naturalists and historians, while using such of their facts as seem to suit the writer's purpose." "The Evolution of Woman," *Chicago Daily Tribune*, March 31, 1894, p. 10.

39. H.L. Mencken, *In Defense of Women* (New York: Alfred A. Knopf, 1918. Reprint, New York: Octagon Books, 1977), 60. Page citations are to the reprint edition. He discussed Gamble on pages 60–64, concluding, "But with the world what it is, it must be obvious that their [woman's] display of finery—to say nothing of their display of epidermis—has the conscious purpose of attracting the masculine eye. A normal woman, indeed, never so much as buys a pair of shoes or has her teeth plugged without considering, in the back of her mind, the effect upon some unsuspecting candidate for her 'reluctant' affections" (64), a sentiment with which Gamble would have likely agreed.

40. "A Fearless Assault on Men," *New York Times*, March 11, 1894, p. 23. For additional reviews of Gamble's book, see "The Evolution of Woman," *Critic*, July 14, 1894, p. 21; "Sex Predominance in Historical Development," *Nation* 58, June 14, 1894, pp. 452–3.

41. "A Fearless Assault on Men."

42. Review of *The Evolution of Woman*, *Popular Science Monthly* 46 (December 1894): 275.

43. Review 1, *Current Literature* 16 (August 1894): 187.

44. "The Evolution of Woman."

45. "Sex Predominance in Historical Development."

46. Jane Johnstone Christie, *The Advance of Woman: From the Earliest Times to the Present* (Philadelphia: J.B. Lippincott, 1912), 8.

47. Alfred Russel Wallace, "Women and Natural Selection," reprinted in *Lucifer the Light-Bearer*, September 15 and October 4, 1894.

48. "Natural Eugenics," *Masses*, September 1913, p. 3.

49. For an extended discussion of Wallace's change of mind regarding sexual selection, see Martin Fichman, *An Elusive Victorian: The Evolution of Alfred Russel Wallace* (Chicago: University of Chicago Press, 2004), ch. 5, "Land Nationalism to Socialism: The Ethics of Politics and the Politics of Ethics." Wallace cited two books as converting him to socialism, Bellamy's *Looking Backward* and Henry George's *Progress and Poverty*. Other Wallace biographers concur that Bellamy inspired Wallace's commitment to socialism and female choice: see Peter Raby, *Alfred Russel Wallace: A Life* (Princeton: Princeton University Press, 2001), 255; and Ross A. Slotten, *The Heretic in Darwin's Court: The Life of Alfred Russel Wallace* (New York: Columbia University Press, 2004), 436–8. See also David Stack, *The First Darwinian Left: Socialism and Darwinism, 1859–1914* (Cheltenham: New Clarion Press, 2003), 29.

50. Matthew Beaumont, introduction to *Looking Backward: 2000–1887* by Edward Bellamy, ed. and with an introduction by Matthew Beaumont, Oxford World Classics Series (New York: Oxford University Press, 2007), vii.

51. Beaumont, introduction to *Looking Backward*, ix.

52. Erich Fromm, introduction to Edward Bellamy, *Looking Backward: 2000–1887* (1888; New York: Signet Classics Edition, 1960).

53. Fichman, *An Elusive Victorian*, 190, 222–3, 262.

54. Alfred Russel Wallace, *My Life: A Record of Events and Opinions*, vol. 2. Facsimile reprint (Westmead: Gregg International Publishers, 1969), 389. Quoted in Fichman, *Elusive Victorian*, 271, in text citation.

55. Sylvia E. Bowman, *Edward Bellamy*, Twayne's United States Authors Series, ed. David J. Nordloh (Boston: Twayne Publishers, 1986), 102.

56. Sylvia Strauss, "Gender, Class, and Race in Utopia," in *Looking Backward: 1988–1888: Essays on Edward Bellamy*, ed. Daphne Patai (Amherst: University of Massachusetts Press, 1988), 68. See also Sylvia Strauss, *Traitors to the Masculine Cause: The Men's Campaign for Women's Rights*, Contributions in Women's Studies, number 35 (Westport, CT: Greenwood Press, 1982).

57. Bowman, *Edward Bellamy*, 102. Bowman does not distinguish between natural and sexual selection in her discussion of the impact of Darwinian theory on Bellamy, but it is clear from his use of the term that Bellamy was referencing sexual selection by female choice as described in *The Descent of Man*.

58. Bowman, *Edward Bellamy*, 104. She is quoting Bellamy, *Looking Backward*, 219.

59. Arthur Lipow, *Authoritarian Socialism in America: Edward Bellamy and the Nationalist Movement* (Berkeley: University of California Press, 1982), 51.

60. Strauss, "Gender, Class, and Race," 86–87.

61. Lester Frank Ward, "Our Better Halves," *Forum* 6 (November 1888): 266.

62. Correa Moylan Walsh, *Feminism* (New York: Sturgis and Walton Company, 1917), 149–50.

63. Walsh, *Feminism*, 158.

64. For another study of socialist marriage reform proposals advanced by Lester Frank Ward, Edward Bellamy, Charlotte Perkins Gilman, Thorstein Veblen, and Theodore Dreiser, see Sondra R. Herman, "Loving Courtship or the Marriage Market? The Ideal and Its Critics, 1871–1911," *American Quarterly* 25 (May 1973): 235–52.

65. Judith Allen, *The Feminism of Charlotte Perkins Gilman: Sexualities, Histories, Progressivism* (Chicago: University of Chicago Press, 2009), ch. 3, "Gynaecocracy and Androcracy."

66. Allen, *Feminism of*, 352.

67. Charlotte Perkins Gilman, *The Man-Made World Or, Our Androcentric Culture* (New York: Charlton Company, 1911), 52.

68. Charles R. Darwin, *Living Cirripedia, A Monograph on the Sub-Class Cirripedia, with Figures of All the Species. The Lepadidæ; or, Pedunculated Cirripedes*, vol. 1 (London: Ray Society, 1851); Charles R. Darwin, *Living Cirripedia, The Balanidæ, (or sessile cirripedes); the Verrucidæ*, vol. 2 (London: Ray Society, 1854). Darwin, *Descent*, 242.

69. Charlotte Perkins Gilman, "Improving on Nature," *Forerunner* 3 (July 1912): 174–6. Gilman also discussed the importance of female choice in *Women and Economics*, 92. The *New York Times* ran an article explaining Gilman's views on female choice, "Charlotte Perkins Gilman Puts Man on the Grill," *New York Times*, January 15, 1911, p. 14.

70. Allen, *The Feminism of*, 353.

71. For an extended study of Bellamy's and Gilman's uses of Darwinian sexual selection, see Kimberly A. Hamlin, "Sexual Selection and the Economics of Marriage: 'Female Choice' in the Writings of Edward Bellamy and Charlotte Perkins Gilman," in *America's Darwin: Darwinian Theory and U.S. Culture, 1859–present*, ed. Lydia Fisher and Tina Gianquitto (Athens, GA: University of Georgia Press, forthcoming).

72. On the ship home to America, the narrator inquired about the young women passengers and was shocked to learn that Miss Elwell, "quite the prettiest woman on board," was a civil engineer. Charlotte Perkins Gilman, *Moving the Mountain*, reprinted in *Charlotte Perkins Gilman's Utopian Novels*, ed. and with an introduction by Minna Doskow (Madison: Associated University Presses, 1999), 45. Page citations are to the reprint edition.

73. Gilman, *Moving the Mountain*, 150.

74. Gilman, *Moving the Mountain*, 46.

75. Gilman, *Moving the Mountain*, 58.

76. Jean H. Baker, *Margaret Sanger: A Life of Passion* (New York: Hill and Wang, 2011), 10.

77. Margaret Sanger, *Woman and the New Race*, with a preface by Havelock Ellis (New York: Blue Ribbon Books, 1920). Baker describes the sales history of the book in *Margaret Sanger*, 160–62.

78. This incident is recounted in Margaret Sanger *An Autobiography* (New York: W.W. Norton and Company, 1938), 20–21.

79. Sanger, *Autobiography*, 107–108.

80. Sanger, *Autobiography*, 108–109. According to Gilman biographer Judith Allen, Gilman did not initially advocate birth control because she feared that it would be used as a "masculinist tool" for coercing women to have frequent sex on men's terms: "for Gilman, birth control mitigated consequences rather than resolving the problem of androcentric sex relations." Allen, *Feminism of*, 321. When Gilman refused to subscribe to the *Woman Rebel*, Sanger publicly chastised Gilman in the *New York Times* as "a conservative and reactionary," but the two later reconciled as Gilman's ideas on birth control evolved. Allen, *Feminism of*, 322–3.

81. Mark Pittenger, *American Socialists and Evolutionary Thought, 1870–1920*, History of American Thought and Culture Series, ed. Paul S. Boyer (Madison: University of Wisconsin Press, 1993). See also, Mark Pittenger, "Evolution, Women's Nature, and American Feminist Socialism, 1900–1915," *Radical History Review* 36 (1986): 47–61. For the classic study of women in American socialism, see Mari Jo Buhle, *Women and American Socialism, 1870–1920* (Urbana: University of Illinois Press, 1983).

82. "Notes," *Socialist Woman* 1, no.1 (June 1907): 4. *The Descent of Man* was included on another reading list published later that year as, "Books on the Woman Question," *Socialist Woman* 1, no. 5 (October 1907): 6. The study group was discussed in "Socialist Women of Chicago," *Socialist Woman* 1, no. 5 (October 1907): 6.

83. Sara Kingsbury, "The Lady-like Woman: Her Place in Nature," *Socialist Woman* 2, no.3 (August 1908): 9.

84. See, for example, Lena Morrow Lewis, "The Sex and Woman Questions," *Masses*, December 7, 1911; "Natural Eugenics," *Masses*, September 1913. The *Masses* frequently published cartoons mocking race suicide by presenting poor women surrounded by starving children and bills; see, for example, Arthur Young, "Hell on Earth," *Masses*, March 1915; and "Race Suicide Alarmists: Congratulations!" *Masses*, February 1914. The *Masses* also recommended Havelock Ellis's articles on birth control, published in Bernarr MacFadden's *Physical Culture*; "Birth Control," *Masses*, August 1915, p. 21.

85. Pittenger, "Evolution, Women's Nature, and American Feminist Socialism," 58.

86. For more on this time in Sanger's life, see Baker, *Margaret Sanger*, 47–74; see also Chesler, *Woman of Valor*, ch. 3, "Seeds of Rebellion."

87. Baker, *Margaret Sanger*, 54.

88. Chesler, *Woman of Valor*, 58.

89. Chesler, *Woman of Valor*, 62–65; Baker, *Margaret Sanger*, 51–52.

90. See, for example, Ethel Cole, "Open Discussion," *Woman Rebel* 3 (May 1914): 18; Alixe Humane, "Ellen Key's Ideal of Woman," *Woman Rebel* 3 (May 1914): 18; Alice Groff, "The Marriage Bed," *Woman Rebel* 4 (June 1914): 39.

91. Gordon, *Moral Property*, 57. She analyzes "voluntary motherhood" on pages 55–70.

92. Baker describes these 1920s trips in chapter 7, "Voyages," in *Margaret Sanger*.

93. For a history of the Comstock laws, see Helen Lefkowitz Horowitz, *Rereading Sex: Battles over Sexual Knowledge and Suppression in Nineteenth-Century America* (New York: Alfred A. Knopf, 2002).

94. Sanger, *Autobiography*, 128.

95. Copies of the early Malthusian League pamphlets are located in the Margaret Sanger Papers, Unfilmed Collection, Sophia Smith Collection, Smith College, Northampton, Massachusetts (SSC).

96. See, for example, 1926 Le Brasseur Company's list of Surgical Rubber Specialties, Margaret Sanger Papers, box 10, folder 1, SSC.

97. C.R. Drysdale, ed. *Medical Opinions on the Population Question* (London: Geo. Standring, 1901), 1. Copy residing at the Margaret Sanger Papers, series III, box 29, folder 1, SSC.

98. C.V. Drysdale, President of the Malthusian League, "The Scientific Basis of Birth Control," reprinted from *Science and Society*, vol. 1 (London, 1937), 16. Margaret Sanger, Unfilmed Collection, box 82, folder 1, SSC.

99. Sanger, *Autobiography*, 127. For a history of the Malthusian League, see Rosanna Ledbetter, *A History of the Malthusian League, 1877–1927* (Columbus: Ohio State University Press, 1976).

100. C.V. Drysdale, "The Scientific Basis of Birth Control," 16, 9.

101. Reprinted excerpt from Geoffrey West, *The Life of Annie Besant* (London: Howe, 1929), Margaret Sanger Papers, series III, box 26, SSC. See also Anne Taylor, *Annie Besant, A Biography* (New York: Oxford, 1992), 116–7.

102. Charles R. Darwin to Charles Bradlaugh, June 6, 1877, letter 10988, Charles Darwin Correspondence Project, Cambridge University, http://www.darwinproject.ac.uk/entry-10988. This is a prepublication stage of the letter, and the Darwin Correspondence Project cannot be held responsible for any errors of transcription remaining. Special thanks to the editors of the Darwin Correspondence Project for access to this unpublished material.

103. See, for example, *The Malthusian Handbook* (London: W.H. Reynolds, 1900), which extensively quoted the *Origin of Species*, copy residing at the Rare Book, Manuscript, and Special Collections Library, Duke University.

104. Sanger, *Autobiography*, 127.

105. Sanger, *Autobiography*, 125.

106. Jean Baker, *Margaret Sanger*, 136. Vincent Brome, *Havelock Ellis, Philosopher of Sex: A Biography* (Boston: Routledge & Kegan Paul, 1979), 247.

107. Sanger, *Autobiography*, 135.

108. Sanger, *Autobiography*, 137.

109. Sanger, *Autobiography*, 141.

110. Transcript of interview with Grant Sanger, by Jacqueline Van Voris, March 28, 1977, Margaret Sanger Papers, series 1, box, 20, folder 8, SSC. Thanks to Alex Sanger for permission to publish material from this interview.

111. Chesler, *Woman of Valor*, 112.

112. Havelock Ellis, *Sexual Selection in Man*, vol. 4 in *Studies in the Psychology of Sex* (Philadelphia: F. A. Davis Company, 1918), v–vi.

113. For a study of the links between evolutionary science and the emergence of sexology as a field, see Kimberly A. Hamlin, "The Birds and the Bees: Darwin's Evolutionary Approach to Human Sexuality," in *Darwin in Atlantic Cultures: Evolutionary Visions of Race, Gender and Sexuality*, ed. Jeannette Eileen Jones and Patrick Sharp

(New York: Routledge Press, 2010). Vern Bullough, too, contends that Darwin's theory of sexual selection "proved a strong impetus for studies in sexuality" and was a "major factor in removing some of the stigma from studying sex." Bullough, *Science in the Bedroom: A History of Sex Research* (New York: Basic Books, 1994), 5. Many pioneering works about homosexuality also reference Darwin and *The Descent of Man*. See, for example, the works of Edward Carpenter (especially *The Intermediate Sex*, 1908).

114. John Allen Godfrey, *The Science of Sex: An Essay toward the Practical Solution of the Sex Problem* (London: University Press, 1901), 4–5. Copy residing at the Kinsey Institute for Sex, Gender, and Reproduction, Inc., Bloomington, Indiana.

115. Godfrey, *The Science of Sex*, 7–8.

116. Patrick Geddes and J. Arthur Thomson, *The Evolution of Sex*, The Contemporary Science Series, ed. Havelock Ellis (New York: Scribner and Welford, 1890), 5.

117. See, for example, Cynthia Eagle Russett's discussion of Geddes and Thomson in *Sexual Science: The Victorian Construction of Womanhood* (Cambridge: Harvard University Press, 1989), 89–92; and Jill Conway, "Stereotypes of Femininity in a Theory of Sexual Evolution," in *Suffer and Be Still: Women in the Victorian Age*, ed. Martha Vicinus (Bloomington: Indiana University Press, 1972): 140–54.

118. Patrick Geddes and J. Arthur Thomson, *Sex* (New York: Henry Holt, 1914), 19.

119. Geddes and Thomson, *The Evolution of Sex*, 292.

120. Geddes and Thomson, *The Evolution of Sex*, 295–7.

121. Geddes and Thomson, *Sex*, 157.

122. Edith How-Martyn, *The Birth Control Movement in England* (London: John Bale Sons and Danielson, 1930), 11. This book then describes how, following Sanger's 1914 visit to England, the American and English birth control movements developed in tandem. Margaret Sanger Papers, series V, box 61, folder, 2, SSC.

123. Havelock Ellis, "The Changing Status of Women," *Westminster Review* 128 (October 1887).

124. Havelock Ellis, "The Love Rights of Women" (New York: Birth Control Review, 1921), 7. Copy residing in the Margaret Sanger Papers, Unfilmed Collection, box 29, folder 6, SSC.

125. Baker, *Margaret Sanger*, 93. For a study of what later historians called "sex positive" feminism in the nineteenth century, see Ellen Carol Dubois and Linda Gordon, "Seeking Ecstasy on the Battlefield: Danger and Pleasure in Nineteenth-Century Feminist Sexual Thought," *Feminist Studies* 9 (Spring 1983): 7–25. See also, Dubois, "Feminism and Free Love," an unpublished essay that she posted on H-Net as part of a scholarly forum on the topic, Humanities and Social Sciences Online, 2001. http://www.h-net.org/~women/papers/freelove.html.

126. Sheila Rowbotham and Jeffrey Weeks, *Socialism and the New Life: The Personal and Sexual Politics of Edward Carpenter and Havelock Ellis* (London: Pluto Press, 1977), 172–3. For more on Ellis's socialist beliefs, see Phyllis Grosskurth, *Havelock Ellis: A Biography* (New York: Alfred A. Knopf, 1980) 61, 69; and Rowbotham and Weeks, *Socialism and the New Life*, 146–7.

127. Baker, *Margaret Sanger*, 161.

128. Jesse Battan, "Sexual Selection and Social Revolution: Anarchist Eugenics and Radical Darwinism in the United States, 1850–1910," in *Darwin in Atlantic Cultures:*

*Evolutionary Visions of Race, Gender, and Sexuality*, ed. Jeannette Eileen Jones and Patrick B. Sharp (New York: Routledge, 2010), 35. Free love advocates also promoted female choice as a reform strategy at the end of the nineteenth century, most notably through the periodical *Lucifer the Light-Bearer* published by Moses Harmon. Free love advocates believed that marriage laws should be rewritten, or undone, to enable women to follow their natural passion and select for fathers the men to whom they were most attracted, whether or not they were married. But, as these reformers turned toward the increasingly institutionalized eugenics movement after 1900, arguments grounded in women's rights fell by the wayside. For more on this gendered split in the sex reform community, see Joanne E. Passet, *Sex Radicals and the Quest for Women's Equality*, Women in American History Series, ed. Anne Firor Scott, Nancy A. Hewitt, and Stephanie Shaw (Urbana: University of Illinois Press, 2003), 166–71. For additional studies of free love and other "sex radical" reforms, see Hal D. Sears, *The Sex Radicals: Free Love in High Victorian America* (Lawrence: Regents Press of Kansas, 1977); David Stack, *The First Darwinian Left: Socialism and Darwinism, 1859–1914* (Cheltenham: New Clarion Press, 2003); and Susan Rensing, "Feminist Eugenics: From Free Love to Birth Control, 1880–1930" (Ph.D. diss., University of Minnesota, 2006).

129. For histories of eugenics, see Daniel Kevles, *In the Name of Eugenics: Genetics and the Uses of Human Heredity* (New York: Alfred A. Knopf, 1985). Kevles discusses the impact of Darwinian theory on the movement but focuses on natural selection; he does not mention sexual selection or *The Descent of Man*. See also Wendy Kline, *Building a Better Race: Gender, Sexuality, and Eugenics from the Turn of Century to the Baby Boom* (Berkeley: University of California Press, 2001); Diane B. Paul, *Controlling Human Heredity, 1865 to the Present* (Atlantic Highlands, NJ: Humanities Press, 1995). Paul presents an overview of eugenic ideology and discusses *The Descent of Man* in chapter 3, "Evolutionary Anxieties." For a focus on the West Coast eugenics movement, see Alexandra Minna Stern, *Eugenic Nation: Faults and Frontiers of Better Breeding in Modern America* (Berkeley: University of California Press, 2005). For studies of eugenics in relation to larger discussions of reproduction, see Gordon, *Moral Property*, 80–85; Laura Lovett, *Conceiving the Future: Pronatalism, Reproduction, and the Family in the United States, 1890–1938* (Chapel Hill: University of North Carolina Press, 2007); Dana Seitler, "Unnatural Selection: Mothers, Eugenic Feminism, and Charlotte Perkins Gilman's Regeneration Narratives," *American Quarterly* 55 (March 2003): 61–88; and Dana Seitler, *Atavistic Tendencies: The Culture of Science in American Modernity* (Minneapolis: University of Minnesota Press, 2008).

130. For more on the relationship between Sanger and the organized eugenics movement, especially during the height of its popularity in the United States in the 1920s, see Chesler, *Woman of Valor*, 215–7.

131. C.V. Drysdale, "Neo-Malthusiansim and Eugenics," pamphlet, Malthusian League, London, 1912. Margaret Sanger Papers, Unfilmed Collection, box 49, folder 1, page 11, SSC.

132. Drysdale, "Neo-Malthusiansim and Eugenics," 14.

133. Baker, *Margaret Sanger*, 164. Baker also clarifies that while Sanger and the American eugenicists did share some common goals, they were never allied.

134. Baker, *Margaret Sanger*, 161.

135. Sanger, *Woman and the New Race*, 36–37.

136. Sanger, *Woman and the New Race*, 44. Some later critics—especially modern opponents of Planned Parenthood—have depicted Sanger's efforts to spread birth control information to African Americans as a nefarious attempt to curtail black population growth, but Sanger biographers clarify that, to the contrary, Sanger simply wanted to share with everyone the birth control information that middle-class, white women already had. Sanger's biographers further contend that her racial views were ahead of her time. According to Jean Baker, she opposed segregation and endeavored to open reproductive health care clinics for African Americans—staffed, ideally, by African American, not white, physicians. Baker, *Margaret Sanger*, 199, 251–53. Both Baker and Ellen Chesler also describe Sanger's efforts to establish a clinic in Harlem in the 1920s and a later network of clinics throughout the South for African Americans as testaments to her concern that African Americans had been unfairly excluded from other health and welfare programs. Chesler, *Woman of Valor*, 295–7 and 388–9; Baker, *Margaret Sanger*, 200–202.

137. Sanger, *Woman and the New Race*, 46.

138. Ellis, preface to *Woman and the New Race*, viii.

139. Sanger, *Woman and the New Race*, 227–9.

140. For histories of the birth control pill, see Elizabeth Siegel Watkins, *On the Pill: A Social History of Oral Contraceptives, 1950–1970* (Baltimore: Johns Hopkins University Press, 1998); Andrea Tone, *Controlling Reproduction: An American History* (Wilmington, DE: Scholarly Resources, Inc., 1997); and Elaine Tyler May, *America and the Pill: A History of Promise, Peril, and Liberation* (New York: Basic Books, 2010).

141. By the late 1930s, Gilman did publicly advocate birth control and support Sanger. The two first met in 1931. At Sanger's urging, Gilman testified before the House Ways and Means Committee on behalf of removing birth control from the category of obscene and allowing doctors to talk about it with their patients and prescribe devices, and she also spoke at the 1934 convention of Sanger's Birth Control League. For Gilman's evolving position on birth control, see Allen, *Feminism of*, 314–23.

142. This is the quote with which the introduction began. Floyd Dell, *Women as World Builders: Studies in Modern Feminism* (Chicago: Forbes and Company, 1913), 44.

CONCLUSION

1. Since the rise of sociobiology in the 1980s and, more recently, evolutionary psychology, feminist scholars have repeatedly debunked the claims of those who argue that evolution is deterministic and patriarchal, often by using evolutionary science themselves. For examples of modern feminist Darwinians working in this vein, see Joan Roughgarden, Jane Lancaster, Barbara Smuts, Sarah Blaffer Hrdy, Marlene Zuk, and Patricia Adair Gowaty. Several additional examples are discussed in the review article by Anne Fausto Sterling, Patricia Adair Gowaty, and Marlene Zuk, "Evolutionary Psychology and Darwinian Feminism," *Feminist Studies* 23 (Summer 1997): 403–17. For critiques of evolutionary psychology, see Roger N. Lancaster, *The Trouble with Nature: Sex and Science in Popular Culture* (Berkeley: University of California Press, 2003); Hilary Rose and Steven Rose, eds., *Alas Poor Darwin: Arguments against Evolution-*

*ary Psychology* (New York: Harmony Books, 2000); Martha McCaughey, *The Caveman Mystique: Pop-Darwinism and the Debates over Sex, Violence and Science* (New York: Routlegde, 2008); Patricia Adair Gowaty, ed., *Feminism and Evolutionary Biology: Boundaries, Intersections, and Frontiers* (New York: Chapman and Hall, 1997); Patricia Adair Gowaty, "Sexual Natures: How Feminism Changed Evolutionary Biology," *Signs* 28 (Spring 2003): 901–23; and the special issue of *Hypatia* on feminist critiques of evolutionary psychology, vol. 27 (Winter 2012).

Page numbers followed by *f* indicate a figure.